LANZHOU LECTURES ON HENSTOCK INTEGRATION

SERIES IN REAL ANALYSIS

Also in the series

Volume 1 Lectures on the Theory of Integration
R. Henstock

Series in Real Analysis Volume 2

LANZHOU LECTURES ON HENSTOCK INTEGRATION

Lee Peng-Yee
National University of Singapore

World Scientific
Singapore • New Jersey • London • Hong Kong

Published by

World Scientific Publishing Co. Pte. Ltd.,
P O Box 128, Farrer Road, Singapore 9128
USA office: 687 Hartwell Street, Teaneck, NJ 07666
UK office: 73 Lynton Mead, Totteridge, London N20 8DH

LANZHOU LECTURES ON HENSTOCK INTEGRATION

ISBN 9971-50-891-5
9971-50-892-3 (pbk)

Printed in Singapore by JBW Printers & Binders Pte. Ltd.

CONTENTS

PREFACE

This is an introductory book on Henstock integration. It is also known as the Kurzweil integral or the generalized Riemann integral. We hope the present book will serve the following three purposes: to report on some recent advances in the theory; to serve as a textbook for a course in the subject; and to stimulate further research in this exciting field. The present book overlaps very little with a recent book by Henstock (Series in Real Analysis Volume 1, World Scientific 1988).

The first draft of these lecture notes, containing the first three chapters, was a written record of what the author said in Lanzhou, China, in October/November 1985 during his visit to Northwest Normal University there. This is the third draft. The second has a limited circulation of only 100 copies. Chapter 4 contains the work of Ding Chuan Song and his Lanzhou school. The topics in Chapter 5 were chosen for their importance and potential for further development.

The Henstock integral is designed to integrate highly oscillatory functions which the Lebesgue integral fails to do. It is known as nonabsolute integration and is a powerful tool. It is well-known that the Henstock integral includes the Riemann, improper Riemann, Lebesgue and Newton integrals. Though such an integral was defined by Denjoy in 1912 and also by Perron in 1914, it was difficult to handle using their definitions. But with the Riemann-type definition introduced recently by Henstock and also independently by Kurzweil, the definition is now simple and furthermore the proof involving the integral also turns out to be easy. For example, a great part of Chapters 1 and 2 can be understood with a prerequisite of advanced calculus.

A key result in the theory is the controlled convergence theorem. It is in a sense the best convergence theorem possible for the Henstock integral, just as the dominated convergence theorem is the best possible for the Lebesgue integral. We give three proofs to the theorem and some applications, through which we introduce some of the standard technique in the theory. The presentation in Chapters 1 and 2 are relatively self-contained and elementary. Chapters 3 and 5 should be read only after having gone through the first two chapters. Fubini's theorem is not included here since it is already expertly covered in Henstock's recent book.

There is active research recently in nonabsolute integration. The present interest is its generalization to n-dimensional euclidean spaces and applications to ordinary differential equations and trigonometric series. After all, nonabsolute integration was developed originally to solve problems in the above-mentioned areas. It is therefore natural to believe that it holds great potential for applications to these fields and perhaps many others. How useful the integral is depends on how much we can apply it to other fields.

After working on the Denjoy-Perron integral for some years, we begin to realize that it is worthwhile to study the integral in the version of Henstock-Kurzweil even if our main interest is only on the absolutely integrable functions. Chapter 1 provides a basic knowledge of the Henstock integral up to the dominated convergence theorem and differentiability properties. There is no shortage of exercises. They can come from the standard course in real analysis or by filling up the gaps in the text such as an alternative proof of the same result but using a different definition of the integral.

The first acknowledgement goes to Professor Bai Guang Bi, the president of Northwest Normal University, for his invitation to give a series of lectures in Lanzhou. The visit initiated the writing of these lecture notes. Many Chinese mathematicians have contributed to the present draft, in particular, Professor Ding Chuan-Song, Mr Lu Shi-Pan, Mr Xu Dong-Fu, Mr Liao Ke-Cheng, and Mr Ma Zhen-Min. Dr Chew Tuan-Seng read through the notes, and suggested various improvements. Thanks are also due to my family (immediate and extended) and colleagues (at home and overseas) for their support and help, without which this book would not have been completed. Miss Sharon Han typed the first draft, Miss Michelle Goh the second, and Miss Tan Bee-Hua the final one. Their assistance is greatly appreciated.

Lee Peng Yee

February 1989

LANZHOU LECTURES ON HENSTOCK INTEGRATION

CHAPTER 1 THE HENSTOCK-KURZWEIL INTEGRAL

1. INTRODUCTION

We may define an integration process in two ways. One is descriptive and another is constructive. For example, Newton defined his integral as an anti-derivative and this is known as a descriptive definition. On the other hand, Riemann defined his integral by first taking the sum and then the limit which is known as a constructive definition. It is well-known that the two integrals do not include each other. We consider in what follows functions defined on a closed bounded interval $[a,b]$. We shall do so throughout the book except some sections in Chapter 5 or otherwise mentioned.

Definition 1.1. A real-valued function f is said to be Newton integrable on $[a,b]$ if there exists a differentiable function F defined on $[a,b]$ such that its derivative $F'(x) = f(x)$ for every x in $[a,b]$. The function F is called the primitive of f on $[a,b]$ and the integral of f on $[a,b]$ is given by $F(b) - F(a)$.

Definition 1.2. A real-valued function f is said to be Riemann integrable to A on $[a,b]$ if for every $\epsilon > 0$ there is a constant $\delta > 0$ such that whenever a division D given by

$$a = x_o < x_1 < \ldots < x_n = b \quad \text{and} \quad \{\xi_1, \xi_2, \ldots, \xi_n\}$$

satisfies $x_{i-1} \le \xi_i \le x_i$ and $x_i - x_{i-1} < \delta$ for all i we have

$$\left| \sum_{i=1}^{n} f(\xi_i)(x_i - x_{i-1}) - A \right| < \epsilon.$$

The integral of f on $[a,b]$ is given by A.

Example 1.3. Let $F(x) = x^2 \sin x^{-2}$ when $x \neq 0$ and $F(0) = 0$. Since F is differentiable at every point, its derivative F' is Newton integrable on $[0,1]$ but not Riemann integrable there. However it is improper Riemann integrable on $(0,1]$ with the singular point at 0.

It is easy to give an example of a function which is Riemann integrable but not Newton integrable, for example, a step function. It is interesting to note that the above derivative F' is not Lebesgue integrable on [0,1]. We shall define the Lebesgue integral later. It is known that the Lebesgue integral includes that of Riemann, therefore the Lebesgue and Newton integrals also do not include each other. Hence we are interested to find an integral that includes both. Denjoy defined such an integral in 1912 and Perron another in 1914. It was almost ten years later that the two integrals were proved to be equivalent. There has been active research on the Denjoy-Perron integral until 1935.

The break-through came in 1957/58 when Henstock and Kurzweil gave independently a Riemann-type definition to the Denjoy integral. Not only that the definition is now easier, but also the proofs using the Henstock–Kurzweil integral are often simpler. In the subsequent years, Henstock and others developed the theory further. Note that if f is Riemann integrable on [a,b], so is its absolute value $|f|$. We say that the Riemann integral is an absolute integral. In the same sense, the Lebesgue integral is also an absolute integral. However the Newton integral and subsequently the Denjoy-Perron-Henstock-Kurzweil integral are nonabsolute. Therefore the Denjoy integral et al are also known as nonabsolute integration. We believe it is still worthwhile to study the Henstock-Kurzweil integral even if our main interest is only on the absolutely integrable functions.

In this book, we present the basic theory of the Henstock integral. In Chapter 1, we shall define the integral, give examples and prove basic properties including the dominated convergence theorem. In Chapter 2, we shall provide several equivalent definitions of the integral. Also, a series of convergence theorems will be proved and they are in a sense better than Lebesgue's dominated convergence theorem. In the next few chapters, we shall consider various aspects of the theory relating to function spaces, Riesz representation theorems for linear and nonlinear functionals (Chapter 3), integration in the plane and some further generalizations (Chapter 5). Other Riemann-type integrals will be presented in Chapter 4.

Almost always, f, g, h, ... denote functions integrable in some sense and F, G, H, ... their primitives. Very often, we write F(u,v) in place of F(v) – F(u). If X is a closed set in [a,b] then we assume known that $(a,b) - X$ can be expressed as the union of a countable number of open intervals. This is essentially a real-line property. Terms beyond advanced calculus are normally defined in the book. Consult the index if necessary. For brevity, a division D given by

$$a = x_o < x_1 < \ldots < x_n = b \quad \text{and} \quad \{\xi_1, \xi_2, \ldots, \xi_n\}$$

is often written $D = \{[u,v];\xi\}$ in which $[u,v]$ represents a typical interval in D, namely, $[x_{i-1},x_i]$ and $([u,v],\xi)$ a typical interval-point pair, namely, $([x_{i-1},x_i],\xi_i)$. Sometimes we write $D = \{[u,v]\}$ if ξ is not involved. Sums over D written

$$\sum f(\xi)(v-u), \quad \sum F(u,v)$$

mean respectively

$$\sum_{i=1}^{n} f(\xi_i)(x_i - x_{i-1}), \quad \sum_{i=1}^{n} F(x_{i-1},x_i).$$

The presentation is elementary up to and including Theorem 5.5. Thereafter we require Vitali's covering theorem and Egoroff's theorem. Lebesgue's theorem on the derivative of a monotone increasing function is used in the proof of Theorem 5.8. Baire's category theorem is used in Theorem 8.12. Both Lebesgue's and Baire's theorems can be avoided if we do not consider the generalized dominated convergence theorem (Theorem 8.12).

2. DEFINITION AND EXAMPLES

Let f be a real-valued function defined on a closed bounded interval $[a,b]$. To define the integral of f on $[a,b]$ using Riemann sums, we may regard each Riemann sum as an estimate of the integral. If f is bounded, then we may partition the interval $[a,b]$ into equal subintervals and compute the Riemann sum as in the definition of the Riemann integral. Suppose f is not bounded and has a singular point at a, i.e., f is not bounded on any subinterval containing a. In order to obtain better estimates, we should use smaller subintervals when closer to a. That is, we should partition the interval $[a,b]$ into unequal subintervals. In other words, we should use a variable δ and not a constant δ as in the definition of the Riemann integral.

To motivate further, let F be a differentiable function defined on [a,b]. Then for each $\xi \in [a,b]$ and for every $\epsilon > 0$ there is a positive function $\delta(\xi) > 0$ depending also on ϵ such that whenever $0 < \xi - u < \delta(\xi)$ and $0 < v - \xi < \delta(\xi)$ we have

$$|F(\xi) - F(u) - f(\xi)(\xi-u)| < \epsilon|\xi-u|,$$

$$|F(v) - F(\xi) - f(\xi)(v-\xi)| < \epsilon|v-\xi|,$$

and consequently

$$|F(v) - F(u) - f(\xi)(v-u)| < \epsilon|v-u|.$$

Suppose given $\delta(\xi)$ we can construct a division D of [a,b] given by

$$a = x_o < x_1 <\ldots< x_n = b \quad \text{and} \quad \{\xi_1,\xi_2,\ldots,\xi_n\}$$

where $0 \le \xi_i - x_{i-1} < \delta(\xi_i)$ and $0 \le x_i - \xi_i < \delta(\xi_i)$ for all i. Then we obtain

$$|F(b) - F(a) - \sum_{i=1}^{n} f(\xi_i)(x_i-x_{i-1})| \le \epsilon(b-a).$$

The above also suggests that we should use a variable δ.

The existence of such divisions is guaranteed by the following lemma.

Lemma 2.1. If $\delta(\xi) > 0$ for $\xi \in [a,b]$, then there is a division D given by

$$a = x_o < x_1 <\ldots< x_n = b \quad \text{and} \quad \{\xi_1,\xi_2,\ldots,\xi_n\}$$

such that $x_{i-1} \le \xi_i \le x_i$ and $[x_{i-1},x_i] \subset (\xi_i-\delta(\xi_i),\ \xi_i+\delta(\xi_i))$ for all i.

This is a direct consequence of the Heine-Borel covering theorem, though it can also be proved directly using continued bisection.

The above discussion leads to the following definition.

Definition 2.2. A real-valued function f is said to be Henstock integrable to A on [a,b] if for every $\epsilon > 0$ there is a function $\delta(\xi) > 0$ such that whenever a division D given by

$$a = x_o < x_1 <\ldots< x_n = b \quad \text{and} \quad \{\xi_1,\xi_2,\ldots,\xi_n\}$$

satisfies $\xi_i \in [x_{i-1},x_i] \subset (\xi_i-\delta(\xi_i),\ \xi_i+\delta(\xi_i))$ for $i = 1,2,\ldots,n$ we have

$$\left|\sum_{i=1}^{n} f(\xi_i)(x_i-x_{i-1}) - A\right| < \epsilon.$$

The integral of f on [a,b] is given by A and A is uniquely determined. The above definition differs from that of Riemann in two ways : $\delta(\xi)$ is no longer a constant and for getting divisions we choose first the points $\xi_1,\xi_2,\ldots,\xi_n$ then $x_o,x_1,\ldots,x_n$ whereas in the case of Riemann we are accustomed to choose first $x_o,x_1,\ldots,x_n$ then $\xi_1,\xi_2,\ldots,\xi_n$. We call ξ_i the associated point of $[x_{i-1},x_i]$ and x_i, $i = 0,1,\ldots,n$, the division points. A division D satisfying the condition in Definition 2.2 is said to be δ-fine. For brevity, we write $D = \{[u,v];\xi\}$ where [u,v] denotes a typical interval in D and ξ is the associated point of [u,v]. If D is δ-fine then $\xi \in [u,v] \subset (\xi-\delta(\xi),\xi+\delta(\xi))$; and vice versa.

Definition 2.2a. A real-valued function f is said to be Henstock integrable to A on [a,b] if for every $\epsilon > 0$ there is a function $\delta(\xi) > 0$ such that for any δ-fine division $D = \{[u,v];\xi\}$ of [a,b] we have

$$\left|\sum f(\xi)(v-u) - A\right| < \epsilon$$

where the sum $\sum$ is understood to be over D.

As usual, we write

$$\int_a^b f(x)dx = A \quad \text{or} \quad \int_a^b f = A.$$

Example 2.3. Every Riemann integrable function on [a,b] is Henstock integrable there. Since a continuous function on [a,b] is Riemann integrable, it is also Henstock integrable.

Example 2.4. Every Newton integrable function on [a,b] is Henstock integrable there. In particular, the derivative defined in Example 1.3 is Henstock integrable on [0,1].

Example 2.5. Let f(x) = 1 when x is rational and 0 when x is irrational. Given $\epsilon > 0$, we label all the rational numbers in [0,1] as r_1, r_2, ... and define $\delta(r_i) = \epsilon\, 2^{-i-1}$ for i = 1, 2, ... and $\delta(\xi) = 1$ otherwise. Then f is Henstock integrable to 0 on [0,1].

We recall that a set of points, X, on the real line is said to be of measure zero if for every $\epsilon > 0$ there is a countable number of open intervals I_1, I_2, ... such that

$$\bigcup_i I_i \supset X \quad \text{and} \quad \sum_i |I_i| < \epsilon$$

where $|I_i|$ denotes the length of I_i. Let f be the characteristic function of X and $X \subset [a,b]$, i.e., $f(x) = 1$ when $x \in X$ and 0 elsewhere in $[a,b]$. Given $\epsilon > 0$, choose I_1, I_2, ... as above and $(\xi - \delta(\xi), \xi + \delta(\xi)) \subset I_i$ for some i when $\xi \in X$ and arbitrary otherwise. Then f is also Henstock integrable to 0 on $[a,b]$.

Example 2.6. We shall verify once again that a continuous function f is Henstock integrable. Let ϵ_n be monotone decreasing to 0 and $\delta_n(\xi) > 0$ such that whenever $|x-\xi| < \delta_n(\xi)$ we have

$$|f(x) - f(\xi)| < \epsilon_n.$$

We may assume $\delta_{n+1}(\xi) \le \delta_n(\xi)$ for all n. Let s_n denote a Riemann sum over a δ_n-fine division. Here s_n, n = 1, 2, ..., are fixed. Take a δ_m-fine division $D = \{[u,v];\xi\}$ and a δ_n-fine division $D' = \{[u',v'];\xi'\}$. If $[u,v] \cap [u',v']$ is non-empty and contains t then

$$\begin{aligned} |f(\xi) - f(\xi')| &\le |f(\xi) - f(t)| + |f(t) - f(\xi')| \\ &< \epsilon_m + \epsilon_n. \end{aligned}$$

It follows that

$$|s_m - s_n| < (\epsilon_m + \epsilon_n)(b-a),$$

and hence $A = \lim_{n\to\infty} s_n$ exists. Therefore given $\epsilon > 0$ there is $\delta_n(\xi) > 0$ with $\epsilon_n < \epsilon$ and $|s_n - A| < \epsilon$ such that over any δ_n-fine division $D = \{[u,v];\xi\}$ we have

$$\begin{aligned} |\textstyle\sum f(\xi)(v-u) - A| &\le |\textstyle\sum f(\xi)(v-u) - s_n| + |s_n - A| \\ &< 2\epsilon(b-a) + \epsilon. \end{aligned}$$

That is, f is Henstock integrable on $[a,b]$. We remark that the above construction of a Cauchy sequence is useful when the integral of the function is not given.

Example 2.7. Let $f(x) = 1/\sqrt{x}$ for $0 < x \le 1$ and $f(0) = 0$. Given $\epsilon > 0$, we shall construct $\delta(\xi)$ so that f is Henstock integrable on $[0,1]$. Consider a division

$$0 = x_o < x_1 < \ldots < x_n = 1 \text{ and } \{\xi_1, \xi_2, \ldots \xi_n\}$$

with $\xi_1 = 0$ and $x_{i-1} \le \xi_i \le x_i$ for $i = 2,\ldots,n$. Note that the primitive of $1/\sqrt{x}$ is $2\sqrt{x}$. Then we can write

$$
\begin{aligned}
\left|2 - \sum_{i=1}^{n} f(\xi_i)(x_i - x_{i-1})\right| &\le \left|2 - (2-\sqrt{x_1})\right| \\
&\quad + \left|\int_{x_1}^{1} dx/\sqrt{x} - \sum_{i=2}^{n} (x_i - x_{i-1})/\sqrt{\xi_i}\right| \\
&\le 2\sqrt{x_1} + \sum_{i=2}^{n} (1/\sqrt{x_{i-1}} - 1/\sqrt{x_i})(x_i - x_{i-1}).
\end{aligned}
$$

We shall prove that the above is less than ϵ for suitable δ-fine divisions. Suppose $\delta(\xi) = c\xi$ for $0 < \xi \le 1$ and $0 < c < 1/2$ so that $\xi_1 = 0$ always. If the above division is δ-fine and $[u,v]$ is a typical interval $[x_{i-1}, x_i]$ in the division with $u \ne 0$ and $u \le \xi \le v$, then $0 < v - u < 2\delta(\xi) \le 2cv$, re-arranging we get $v/u \le 1/(1-2c)$, and finally

$$(v-u)/\sqrt{uv} < 2cv/u \le 2c/(1-2c).$$

Now choose c so that $0 < c < 1/2$ and $2c/(1-2c) \le \epsilon/2$. In addition, put $\delta(0) \le \epsilon^2/16$. Then for the given δ-fine division the above inequality is less than

$$2\sqrt{\delta(0)} + \frac{2c}{1-2c} \sum_{i=2}^{n} (\sqrt{x_i} - \sqrt{x_{i-1}}) < \epsilon.$$

For example, when $0 < \epsilon \le 1$ we may choose $c = \epsilon/6$. Hence the function is Henstock integrable on $[0,1]$.

We may also define the Henstock integral on the real line. Let $\overline{R}$ denote the extended real line, i.e., the set obtained by adding to the real line the conventional points at infinity, namely, $+\infty$ and $-\infty$. Define $\delta(x) > 0$ when $-\infty < x < \infty$, $\delta(-\infty) = A > 0$ and $\delta(+\infty) = B > 0$. A division of $\overline{R}$ given by

$$-\infty < a = x_0 < x_1 < \ldots < x_n = b < +\infty \text{ and } \xi_1, \xi_2, \ldots, \xi_n$$

is said to be δ-fine if $a < -A$, $b > B$ and the bounded interval $[a,b]$ is δ-fine in the usual sense, i.e.,

$$\xi_i - \delta(\xi_i) < x_{i-1} \le \xi_i \le x_i < \xi_i + \delta(\xi_i) \text{ for } i = 1,2,\ldots,n.$$

We may regard $-\infty$ as the associated point of $[-\infty,a]$ and $+\infty$ that of $[b,+\infty]$. Being δ-fine requires that $[-\infty,a] \subset [-\infty,A)$ and $[b,+\infty] \subset (B,+\infty]$. Obviously, given $\delta(x) > 0$ for $x \in \overline{R}$, a δ-fine division of $\overline{R}$ exists.

Then a function f defined on the real line is said to be Henstock integrable to A if for every $\epsilon > 0$ there is a function $\delta(\xi) > 0$ defined on $\bar{R}$ such that for any δ-fine division as given above we have

$$|\sum_{i=1}^{n} f(\xi_i)(x_i - x_{i-1}) - A| < \epsilon.$$

Again the integral A is uniquely determined. Here $f(-\infty)$ and $f(+\infty)$ are assumed to be zero. The contributions from the unbounded intervals to the Riemann sum are also taken to be zero.

Example 2.8. Let $f(x) = x^{-2}$ when $1 \le x < +\infty$. Given $\epsilon > 0$, define $\delta(x) = \epsilon x$ when $1 \le x < +\infty$ and $\delta(+\infty) = 1/\epsilon$. Then for any δ-fine division as given above in the definition with $x_0 = 1$ and $x_n > 1/\epsilon$ we have

$$|\sum_{i=1}^{n} f(\xi_i)(x_i - x_{i-1}) - 1| \le |\sum_{i=1}^{n} \frac{x_i - x_{i-1}}{\xi_i^2} - \sum_{i=1}^{n} (\frac{1}{x_{i-1}} - \frac{1}{x_i})| + \frac{1}{x_n}$$

$$\le \sum_{i=1}^{n} |\frac{x_i x_{i-1}}{\xi_i^2} - 1|(\frac{1}{x_{i-1}} - \frac{1}{x_i}) + \epsilon$$

Considering separately the case when $x_i x_{i-1}/\xi_i^2 - 1$ is positive or negative, we obtain

$$|\frac{x_i x_{i-1}}{\xi_i^2} - 1| \le \frac{x_i - x_{i-1}}{\xi_i} < 2\epsilon$$

Combining the above inequalities, we show that f is Henstock integrable on $[1, +\infty)$.

In what follows, we could have developed the theory of Henstock integration on the real line. In order to keep the presentation simple, we shall confine ourselves to the compact interval [a,b] only. It is not difficult to see how the proof could have been adjusted to work for the case on the real line as well.

3. SIMPLE PROPERTIES

The main result in this section is Henstock's lemma (Theorem 3.7). We shall need it later for proving the convergence theorems.

Theorem 3.1. If f and g are Henstock integrable on [a,b], then so are f + g and αf where α is real. Furthermore,

$$\int_a^b (f+g) = \int_a^b f + \int_a^b g, \quad \int_a^b (\alpha f) = \alpha \int_a^b f.$$

Proof. Let A and B denote respectively the integrals of f and g on [a,b]. Given $\epsilon > 0$, there is a $\delta_1(\xi) > 0$ such that for any δ_1-fine division $D = \{[u,v];\xi\}$ we have

$$|\sum f(\xi)(v-u) - A| < \epsilon/2.$$

Similarly, there is a $\delta_2(\xi) > 0$ such that for any δ_2-fine division $D = \{[u,v];\xi\}$ we have

$$|\sum g(\xi)(v-u) - B| < \epsilon/2.$$

Now put $\delta(\xi) = \min\{\delta_1(\xi),\delta_2(\xi)\}$. Note that any δ-fine division is also δ_1-fine and δ_2-fine. Therefore for any δ-fine division $D = \{[u,v];\xi\}$ we have

$$\begin{aligned}|\sum\{f(\xi) + g(\xi)\}(v-u) - (A+B)| &\le |\sum f(\xi)(v-u) - A| + |\sum g(\xi)(v-u) - B| \\ &< \epsilon.\end{aligned}$$

The proof is complete. The second part is easy.

Theorem 3.2. Let $a < c < b$. If f is Henstock integrable on [a,c] and on [c,b], then so it is on [a,b] and

$$\int_a^b f = \int_a^c f + \int_c^b f.$$

Proof. Let A denote the integral of f on [a,c] and B that of f on [c,b]. Given $\epsilon > 0$, there is a $\delta_1(\xi) > 0$, defined on [a,c], such that for any δ_1-fine division $D = \{[u,v];\xi\}$ of [a,c] we have

$$|\sum f(\xi)(v-u) - A| < \epsilon/2.$$

Similarly, there is a $\delta_2(\xi) > 0$, defined on [c,b], such that for any δ_2-fine division $D = \{[u,v];\xi\}$ of [c,b] we have

$$|\sum f(\xi)(v-u) - B| < \epsilon/2.$$

Define $\delta(\xi) = \min\{\delta_1(\xi),\ c-\xi\}$ when $\xi \in [a,c)$, $\min\{\delta_2(\xi),\ \xi-c\}$ when $\xi \in (c,b]$, and $\min\{\delta_1(c),\ \delta_2(c)\}$ when $\xi = c$. Note that for any δ-fine division D of [a,b], c is always a division point of D. Therefore for any δ-fine division $D = \{[u,v];\xi\}$ of [a,b] with Σ over D, writing

$\Sigma = \Sigma_1 + \Sigma_2$ where Σ_1 is the partial sum over [a,c] and Σ_2 over [c,b] we have

$$|\sum f(\xi)(v-u) - (A+B)| \le |\sum_1 f(\xi)(v-u) - A| + |\sum_2 f(\xi)(v-u) - B|$$
$$< \epsilon.$$

Hence f is Henstock integrable to A + B on [a,b].

Alternativley, let $\chi_{[a,c]}$ denote the characteristic function of [a,c] and $f_1 = f\,\chi_{[a,c]}$. Similarly, let $f_2 = f\chi_{[c,b]}$. Then it follows from Example 2.5 and Theorem 3.1 that

$$\int_a^b f = \int_a^b (f_1+f_2) = \int_a^c f + \int_c^b f.$$

Theorem 3.3. If f is Henstock integrable on [a,b], then so it is on a subinterval [c,d] of [a,b].

First, we observe that the following Cauchy condition holds for the Henstock integral.

Lemma 3.4. A function f is Henstock integrable on [a,b] if and only if for every $\epsilon > 0$ there is a $\delta(\xi) > 0$ such that for any δ-fine divisions $D = \{[u,v];\xi\}$ and $D' = \{[u',v'];\xi'\}$ we have

$$|\sum f(\xi)(v-u) - \sum f(\xi')(v'-u')| < \epsilon$$

where the first sum is over D and the second over D'.

The proof follows easily from the same argument as in Example 2.6.

Proof of Theorem 3.3. Since f is Henstock integrable on [a,b], the Cauchy condition holds. Take any two δ-fine divisions of [c,d], say, D_1 and D_2, and denote by s_1 and s_2 respectively the Riemann sums of f over D_1 and D_2. Similarly, take another δ-fine division D_3 of $[a,c] \cup [d,b]$ and denote by s_3 the corresponding Riemann sum. Then the union $D_1 \cup D_3$ forms a δ-fine division of [a,b]. Here the division points and associated points of $D_1 \cup D_3$ are the union of those from D_1 and D_3. The Riemann sum of f over $D_1 \cup D_3$ is $s_1 + s_3$, and similarly that over $D_2 \cup D_3$ is $s_2 + s_3$. Therefore by the Cauchy condition we have

$$|s_1-s_2| \le |(s_1+s_3) - (s_2+s_3)| < \epsilon.$$

Hence the result follows from Lemma 3.4 with [a,b] replaced by [c,d].

Theorem 3.5. If f(x) = 0 almost everywhere in [a,b], i.e., for every x in [a,b] except perhaps a set X of measure zero, then f is Henstock integrable to 0 on [a,b].

Proof. Note that X is the union of X_i, i = 1, 2, ..., where X_i is a subset of X in which $i - 1 < |f(x)| \le i$ for $x \in X_i$. Each X_i is also of measure zero. Given $\epsilon > 0$, for each i there is a G_i which is the union of a countable number of open intervals with the total length less than $\epsilon 2^{-i} i^{-1}$ and such that $G_i \supset X_i$. Then define $\delta(\xi)$ such that $(\xi - \delta(\xi), \xi + \delta(\xi)) \subset G_i$ for $\xi \in X_i$, i = 1, 2,... , and arbitrarily otherwise. Hence for any δ-fine division $D = \{[u,v];\xi\}$ we have

$$|\sum f(\xi)(v-u)| < \epsilon.$$

The proof is complete.

In what follows, a property is said to hold almost everywhere if it holds everywhere except perhaps in a set of measure zero. Sometimes we say "for almost all x in X" in place of "almost everywhere in X".

Theorem 3.6. If f and g are Henstock integrable on [a,b] and if $f(x) \le g(x)$ for almost all x in [a,b], then

$$\int_a^b f \le \int_a^b g.$$

Proof. In view of Theorem 3.5, we may assume that $f(x) \le g(x)$ for all x. Given $\epsilon > 0$, as in the proof of Theorem 3.1, there is a $\delta(\xi) > 0$ such that for any δ-fine division $D = \{[u,v];\xi\}$ we have

$$|\sum f(\xi)(v-u) - \int_a^b f| < \epsilon,$$

$$|\sum g(\xi)(v-u) - \int_a^b g| < \epsilon.$$

It follows that

$$\int_a^b f - \epsilon < \sum f(\xi)(v-u) \le \sum g(\xi)(v-u) < \int_a^b g + \epsilon.$$

Since ϵ is arbitrary, we have the required inequality,

Theorem 3.7. If f is Henstock integrable on [a,b] with the primitive F, then for every $\epsilon > 0$ there is a $\delta(\xi) > 0$ such that for any δ-fine division $D = \{[u,v];\xi\}$ we have

$$\sum|F(v) - F(u) - f(\xi)(v-u)| < \epsilon.$$

We shall make a few remarks before the proof. From the computational point of view, we may regard $f(\xi)(v-u)$ as an approximation of $F(v) - F(u)$. Then the difference $F(v) - F(u) - f(\xi)(v-u)$ is an error. The definition of the Henstock integral (Definition 2.2) says that the accumulated error is small, whereas Henstock's lemma (Theorem 3.7) says that the absolute error is also small. In fact, the two are equivalent by Theorem 3.7. Another way of putting it is that taking any partial sum Σ_1 of Σ we still have

$$|\textstyle\sum_1\{F(v) - F(u) - f(\xi)(v-u)\}| < \epsilon$$

That is to say, the "selected" error is again small, and indeed it is equivalent to the above two.

Proof of Theorem 3.7. Given $\epsilon > 0$, there is a $\delta(\xi) > 0$ such that for any δ-fine division $D = \{[u,v];\xi\}$ we have

$$|\textstyle\sum\{F(v) - F(u) - f(\xi)(v-u)\}| < \epsilon/4.$$

Let Σ_1 be a partial sum of Σ and E_1 the union of $[u,v]$ from Σ_1. Suppose E_2 is the closure of $[a,b] - E_1$. Then, by Theorem 3.3, f is Henstock integrable on E_2. Thus we can choose a δ-fine division $D_2 = \{[u,v];\xi\}$ of E_2 such that

$$|\textstyle\sum_2\{F(v) - F(u) - f(\xi)(v-u)\}| < \epsilon/4$$

where Σ_2 is over D_2. Now writing $\Sigma_3 = \Sigma_1 + \Sigma_2$ we have

$$\begin{aligned} |\textstyle\sum_1\{F(v) - F(u) - f(\xi)(v-u)\}| &\le |\Sigma_3\{\quad\}| + |\Sigma_2\{\quad\}| \\ &< \epsilon/2. \end{aligned}$$

Consequently, the result follows.

Corollary 3.8. If f is Henstock integrable on $[a,b]$, then its primitive F is continuous on $[a,b]$.

Proof. The continuity follows from Theorem 3.7 and the following inequality

$$|F(t) - F(\xi)| \le |F(t) - F(\xi) - f(\xi)(t-\xi)| + |f(\xi)(t-\xi)|.$$

In fact, the primitive F will satisfy further condition as we shall see in Lemma 6.19.

4. SOME CONVERGENCE THEOREMS

First, we prove the following monotone convergence theorem.

Theorem 4.1. If the following conditions are satisfied:

(i) $f_n(x) \to f(x)$ almost everywhere in $[a,b]$ as $n \to \infty$ where each f_n is Henstock integrable on $[a,b]$;

(ii) $f_1(x) \le f_2(x) \le \ldots$ for almost all $x \in [a,b]$;

(iii) the integrals $F_n(a,b)$ of f_n on $[a,b]$ converges to A as $n \to \infty$,

then f is Henstock integrable to A on $[a,b]$.

Proof. For simplicity, we may assume $f_n(x) \to f(x)$ everywhere as $n \to \infty$. Given $\epsilon > 0$, for every $\xi \in [a,b]$ there is a positive integer $m(\epsilon,\xi)$ such that

$$|f_{m(\epsilon,\xi)}(\xi) - f(\xi)| < \epsilon.$$

Since f_n is Henstock integrable on $[a,b]$, let F_n be the primitive of f_n and write $F_n(u,v) = F_n(v) - F_n(u)$, then by Theorem 3.7 there is a $\delta_n(\xi) > 0$ such that for any δ_n-fine division $D = \{[u,v];\xi\}$ we have

$$\sum|F_n(u,v) - f_n(\xi)(v-u)| < \epsilon\, 2^{-n}.$$

Now put $\delta(\xi) = \delta_{m(\epsilon,\xi)}(\xi)$ for $\xi \in [a,b]$. Take any δ-fine division $D = \{[u,v];\ \xi\}$ and we have

$$\begin{aligned}|\sum f(\xi)(v-u) - A| &\le \sum|f(\xi) - f_{m(\epsilon,\xi)}(\xi)|(v-u) \\ &\quad + \sum|f_{m(\epsilon,\xi)}(\xi)(v-u) - F_{m(\epsilon,\xi)}(u,v)| \\ &\quad + |\sum F_{m(\epsilon,\xi)}(u,v) - A|. \\ &< \epsilon(b-a) + \sum_{n=1}^{\infty} \epsilon 2^{-n} + |\sum F_{m(\epsilon,\xi)}(u,v) - A|.\end{aligned}$$

Therefore it remains to show that the last term of the above inequality is small.

First, we note that the sequence $\{F_n(u,v)\}$ is monotone increasing and convergent to a limit, say, $F(u,v)$. Here $F(a,b) = A$. Further, the number of the associated points ξ in D is finite, and so is the number of those different $m(\epsilon,\xi)$ in the above sum over D. Let p denote the minimum of those $m(\epsilon,\xi)$. Then we have

$$F_p(a,b) = \sum F_p(u,v) \le \sum F_{m(\epsilon,\xi)}(u,v) \le \sum F(u,v) = A.$$

Obviously, we can find m_o such that

$$0 \le A - F_m(a,b) < \epsilon \qquad \text{whenever } m \ge m_o.$$

Therefore when defining $m(\epsilon,\xi)$ we should choose $m(\epsilon,\xi) \ge m_o$. Hence

$$|\sum F_{m(\epsilon,\xi)}(u,v) - A| \le A - F_p(a,b) < \epsilon.$$

The proof is complete.

Next, we need the following lemma for proving the dominated convergence theorem.

Lemma 4.2. If f_1 and f_2 are Henstock integrable on $[a,b]$, and if $g(x) \le f_i(x) \le h(x)$ almost everywhere for $i = 1,2$ where g and h are also Henstock integrable on $[a,b]$, then $\max(f_1,f_2)$ and $\min(f_1,f_2)$ are both Henstock integrable on $[a,b]$.

Proof. First, suppose $g(x) = 0$ for $x \in [a,b]$. Let $F_i(u,v)$ be the integral of f_i on $[u,v]$ for $i = 1,2$ and put

$$F^*(u,v) = \max\{F_1(u,v),\ F_2(u,v)\}.$$

Note that F^* is not additive. When $x < y < z$ we only have

$$F^*(x,z) \le F^*(x,y) + F^*(y,z).$$

Take any division $D = \{[u,v]\}$ of $[a,b]$ and we have

$$0 \le \sum F^*(u,v) \le \int_a^b h(x)dx.$$

Denote by A the supremum of all such $\Sigma F^*(u,v)$. We shall show that A is the integral of $\max(f_1,f_2)$ on $[a,b]$.

Given $\epsilon > 0$, there is a $\delta(\xi) > 0$ such that for any δ-fine division $D = \{[u,v];\xi\}$ of $[a,b]$ we have

$$\sum|f_i(\xi)(v-u) - F_i(u,v)| < \epsilon \qquad i = 1,2.$$

Now put

$$\chi_i(x,y) = \sup \sum|f_i(\xi)(v-u) - F_i(u,v)|, \qquad i = 1,2,$$

where the supremum is over all δ-fine divisions $D = \{[u,v];\xi\}$ of $[x,y]$. Note that

$$\chi_i(x,y) + \chi_i(y,z) \le \chi_i(x,z) \quad \text{for } x < y < z, \text{ and } \chi_i(a,b) \le \epsilon.$$

For any δ-fine division $D = \{[u,v];\xi\}$ of $[a,b]$ we have

$$f_i(\xi)(v-u) \le F^*(u,v) + \chi_1(u,v) + \chi_2(u,v),\ i = 1,2.$$

Consequently, writing $f = \max(f_1,f_2)$ we have

$$f(\xi)(v-u) \le F^*(u,v) + \chi_1(u,v) + \chi_2(u,v).$$

Similarly, we also have

$$F^*(u,v) - \chi_1(u,v) - \chi_2(u,v) \le f(\xi)(v-u).$$

Combining the above we obtain

$$|\sum\{f(\xi)(v-u) - F^*(u,v)\}| \le 2\epsilon.$$

Finally, fix a division $D_1 = \{[u,v]\}$ such that its corresponding sum

$$\sum_1 F^*(u,v) > A - \epsilon.$$

Modify $\delta(\xi)$ in such a way that if D is δ-fine then it is finer than D_1, i.e., any subinterval of D is included in some subinterval of D_1. For any modified δ-fine division $D = \{[u,v];\xi\}$ we have

$$0 \le A - \sum F^*(u,v) \le A - \sum_1 F^*(u,v) < \epsilon.$$

Applying the above inequalities we obtain

$$|\sum f(\xi)(v-u) - A| < 3\epsilon.$$

Hence we have proved the first part for g being a zero function.

In general, we consider $0 \le \max(f_1-g,\ f_2-g) \le h-g$ and show that $\max(f_1-g, f_2-g)$ is Henstock integrable on [a,b]. Since

$$\max(f_1-g,\ f_2-g) = \max(f_1,f_2) - g$$

and g is Henstock integrable on [a,b], so is $\max(f_1,f_2)$. The second part follows from the fact that $\min(f_1,f_2) = -\max(-f_1,-f_2)$.

Note that the condition $g \le f_i \le h$ for $i = 1, 2$ in Lemma 4.2 cannot be omitted. Otherwise the Henstock integrablity of f would imply that of $\max(f,-f) = |f|$ which is incorrect. The given condition is used to prove only the boundedness of $\sum F^*(u,v)$. Hence we have also proved the following

Lemma 4.2a. If f_1 and f_2 are Henstock integrable on [a.b] and if their primitives F_1 and F_2 are both of bounded variation on [a,b], then $\max(f_1,f_2)$ and $\min(f_1,f_2)$ are Henstock integrable on [a,b].

We recall that a function F is said to be of bounded variation on [a,b] if the following total variation is finite:

$$V(F;[a,b]) = \sup \sum |\ F(v) - F(u)|$$

where the supremum is over all divisions $D = \{[u,v]\}$ of [a,b] and $\sum$ is over D.

Now we state and prove the dominated convergence theorem.

Theorem 4.3. If the following conditions are satisfied:

(i) $f_n(x) \to f(x)$ almost everywhere in [a,b] as $n \to \infty$ where each f_n is Henstock integrable on [a,b];

(ii) $g(x) \le f_n(x) \le h(x)$ for almost all x in [a,b] and all n where g and h are also Henstock integrable on [a,b],
then f is Henstock integrable on [a,b] and

$$\int_a^b f_n \to \int_a^b f \qquad \text{as } n \to \infty.$$

Proof. By Lemma 4.2, the function $\min\{f_n;\ i \le n \le j\}$ is Henstock integrable on [a,b]. Denote it by f_j^* for $j = i, i+1, i+2, \ldots$. Then the sequence $-f_i^*, -f_{i+1}^*, \ldots$ is monotone increasing and their integrals are bounded above. By the monotone convergence theorem (Theorem 4.1), the limit function $\inf\{f_n;\ n \ge i\}$ is Henstock integrable on [a,b].

Similarly, we can show that $\sup\{f_n;\ n \ge i\}$ is also Henstock integrable on [a,b]. Then we have

$$\int_a^b (\inf_{n \ge i} f_n) \le \inf_{n \ge i} \int_a^b f_n \le \sup_{n \ge i} \int_a^b f_n \le \int_a^b (\sup_{n \ge i} f_n).$$

It is well-known that $f_n(x) \to f(x)$ as $n \to \infty$ if and only if

$$\lim_{i \to \infty} \{\inf_{n \ge i} f_n(x)\} = f(x) = \lim_{i \to \infty} \{\sup_{n \ge i} f_n(x)\}.$$

Apply the monotone convergence theorem again to the sequence $\inf\{f_n;\ n \ge i\}$ for $i = 1, 2, \ldots$ and we obtain that f is Henstock integrable on [a,b]. Consequently,

$$\int_a^b f \le \lim_{i \to \infty} (\inf_{n \ge i} \int_a^b f_n) \le \lim_{i \to \infty} (\sup_{n \ge i} \int_a^b f_n) \le \int_a^b f,$$

and the result follows.

We recall that given a sequence $\{a_n\}$ the lower limit of the sequence is defined by

$$\liminf_{n \to \infty} a_n = \lim_{i \to \infty} \{\inf_{n \ge i} a_n\}.$$

Similarly, we may define the upper limit.

Corollary 4.4 Let f_n be non-negative and Henstock integrable on [a,b] with $f_n(x) \to f(x)$ almost everywhere in [a,b] as $n \to \infty$. If the sequence of the integrals of $f_1, f_2, \ldots$ is bounded, then f is Henstock integrable on [a,b] and

$$\int_a^b f(x)dx \le \liminf_{n\to\infty} \int_a^b f_n(x)dx.$$

This is known as Fatou's lemma. Note that we do not have the corresponding inequality involving the upper limit.

Corollary 4.5. Theorem 4.3 remains valid with (ii) replaced by:

(iii) $\int_a^b |f_n - f_m| \to 0$ as n, m $\to \infty$.

Proof. In view of (iii), we can choose integers $n(1) < n(2) < \ldots$ such that

$$\int_a^b |f_{n(i)} - f_{n(i+1)}| < 2^{-i} \quad \text{for } i = 1,2, \ldots .$$

Then by the monotone convergence theorem (Theorem 4.1), we show that

$$g(x) = \sum_{i=1}^{\infty} |f_{n(i)}(x) - f_{n(i+1)}(x)|$$

exists almost everywhere and is Henstock integrable on [a,b]. Therefore the sequence $\{f_{n(i)}\}$ is dominated on the right by $f_{n(1)} + g$ and on the left by $f_{n(1)} - g$. By the dominated convergence theorem, f is Henstock integrable on [a,b]. It follows from Fatou's lemma that

$$\int_b^a |f-f_n| \le \liminf_{m\to\infty} \int_b^a |f_m - f_n|.$$

Hence the required condition is satisfied.

This is known as the mean convergence theorem. More convergence theorems will be proved in Chapter 2.

5. ABSOLUTE INTEGRABILITY AND MEASURABILITY

A function f is said to be absolutely Henstock integrable on [a,b] if f and its absolute value |f| are both Henstock integrable on [a,b]. We shall discuss in this section the necessary and sufficient conditions for a function to be absolutely Henstock integrable.

Definition 5.1. A function F is said to be absolutely continuous on [a,b] if for every $\epsilon > 0$ there is a $\eta > 0$ such that for every finite or infinite sequence of non-overlapping intervals $\{[a_i,b_i]\}$ satisfying

$$\sum_i |b_i - a_i| < \eta \quad \text{we have} \quad \sum_i |F(b_i) - F(a_i)| < \epsilon$$

Graphically, a continuous function maps a small interval in the domain into another small interval in the range. An absolutely continuous function maps a collection of small intervals (which are obtained by cutting up a small η interval) in the domain into another collection of small intervals in the range. Note that if we put the collection of small intervals in the range together we obtain a small ϵ interval. It is easy to see that an absolutely continuous function is continuous, and of bounded variation but not conversely. Graphically again, we may add up all the ups and downs of a given function over [a,b] and if the total is finite then it is a function of bounded variation on [a,b].

Theorem 5.2. If f is absolutely Henstock integrable on [a,b], then the primitive F of f is absolutely continuous on [a,b].

To prove the theorem, we need the following lemmas.

Lemma 5.3. If f is absolutely Henstock integrable on [a,b], then the primitive F of f is of bounded variation on [a,b] and

$$\int_a^b |f(x)|dx = V(F;[a,b])$$

where V denotes the total variation of F over [a,b].

Proof. Let A denote the integral of $|f|$ on [a,b]. It follows easily from Theorem 3.6 that $V(F;[a,b]) \le A$. Since both f and $|f|$ are Henstock integrable on [a,b], for every $\epsilon > 0$ there is a $\delta(\xi) > 0$ such that for any δ-fine division $D = \{[u,v];\xi\}$ we have

$$\sum|f(\xi)(v-u) - F(u,v)| < \epsilon,$$

$$|\sum|f(\xi)|(v-u) - A| < \epsilon.$$

As usual, $F(u,v) = F(v) - F(u)$. It follows that

$$\begin{aligned} A &\le |A-\sum|f(\xi)|(v-u)| + \sum|f(\xi)(v-u)-F(u,v)| \\ &\quad + \sum|F(u,v)| \\ &< 2\epsilon + V(F;[a,b]). \end{aligned}$$

Hence $A \le V(F;[a,b]$, and the equality holds.

Given f, a truncated function f^N is defined by $f^N(x) = f(x)$ when

$|f(x)| \le N$, $f^N(x) = N$ when $f(x) > N$, and $f^N(x) = -N$ when $f(x) < -N$.

Lemma 5.4. If f is absolutely Henstock integrable on [a,b], then so is f^N and the primitive of f^N is absolutely continuous on [a.b].

Proof. By Lemma 5.3 and Lemma 4.2a, we show that $g = \min(f,N)$ and $f^N = \max(-N,g)$ are Henstock integrable on [a,b]. Since f^N is bounded, $|f^N| = \max(f^N,-f^N)$ is also Henstock integrable on [a,b].

Let G be the primitive of f^N, and $G(u,v) = G(v) - G(u)$. Then for every $\epsilon > 0$ there is a $\delta(\xi) > 0$ such that for any δ-fine division $D = \{[u,v];\xi\}$ of [a,b] we have

$$\sum|f^N(\xi)(v-u) - G(u,v)| < \epsilon/2.$$

Now choose η such that $2N\eta < \epsilon$. Then for any non-overlapping intervals $[a_k,b_k]$, $k = 1,2,\ldots, n$ with the total length less than η we take a δ-fine division $D = \{[u,v];\xi\}$ of the union of the above intervals and we have

$$\sum_{k=1}^{n} |G(a_k,b_k)| \le \sum|G(u,v) - f^N(\xi)(v-u)| + \sum|f^N(\xi)(v-u)|$$
$$< \epsilon/2 + N\eta < \epsilon.$$

That is, G is absolutely continuous on [a,b].

Proof of Theorem 5.2. Given $\epsilon > 0$, by the dominated convergence theorem we can find a suitable truncated function f^N such that

$$\int_a^b |f-f^N| < \epsilon/2.$$

Let G denote the primitive of f^N. For non-overlapping intervals $[a_k,b_k]$, $k = 1,2,\ldots,n$, we consider

$$\sum_{k=1}^{n} |F(a_k,b_k)| \le \sum_{k=1}^{n} |F(a_k,b_k) - G(a_k,b_k)| + \sum_{k=1}^{n} |G(a_k,b_k)|$$
$$\le \int_a^b |f-f^N| + \sum_{k=1}^{n} |G(a_k,b_k)|.$$

Since G is absolutely continuous on [a,b], then so is F.

Note that proving the results involving the absolute integrability we often divide it into two parts : bounded functions and then the general case. This is standard. Together with Lemma 4.2a, we have proved that for a Henstock integrable function, it is absolutely Henstock integrable if and only if its primitive is absolutely continuous.

Theorem 5.5. If F is absolutely continuous on [a,b] with its derivative $F'(x) = f(x)$ almost everywhere in [a,b], then f is absolutely Henstock integrable on [a,b].

Proof. Suppose $F'(x) = f(x)$ for $x \in [a,b] - X$ where X is of measure zero. Then for $\xi \in [a,b] - X$ and $\epsilon > 0$ there is a $\delta(\xi) > 0$ such that whenever $\xi \in [u,v] \subset (\xi-\delta(\xi), \xi+\delta(\xi))$ and writing $F(u,v) = F(v)-F(u)$ we have

$$|F(u,v) - f(\xi)(v-u)| \leq \epsilon|v-u|.$$

Next put

$$X_i = \{x \in X;\ i - 1 \leq |f(x)| < i\}, \quad i = 1,2, \ldots .$$

Each X_i is also of measure zero. Since F is absolutely continuous, for each i there are a $\eta_i < \epsilon 2^{-i}i^{-1}$ and a sequence of open intervals $I_{i,k}$ for $k = 1,2, \ldots$ such that

$$X_i \subset \bigcup_k I_{i,k}, \quad \sum_k |I_{i,k}| < \eta_i \quad \text{and} \quad \sum_k |F(I_{i,k})| < \epsilon 2^{-i},$$

where $|I_{i,k}|$ denotes the length of $I_{i,k}$ and $F(I) = F(v) - F(u)$ when $I = (u,v)$. For $\xi \in X_i$ and $i = 1,2,\ldots$, we define $\delta(\xi)$ so that $(\xi-\delta(\xi), \xi+\delta(\xi)) \subset I_{i,k}$ for some k. Hence we have defined $\delta(\xi)$ for all $\xi \in [a,b]$.

Take any δ-fine division $D = \{[u,v];\xi\}$ and split the sum $\sum$ over D into two partial sums $\sum_1$ andf $\sum_2$ in which $\xi \notin X$ and $\xi \in X$ respectively. Then we obtain

$$\begin{aligned} |\sum f(\xi)(v-u) - F(a,b)| &\leq \textstyle\sum_1 |f(\xi)(v-u)-F(u,v)| + \sum_2 |F(u,v)| \\ &\quad + \textstyle\sum_2 |f(\xi)(v-u)| \\ &< \epsilon(b-a) + \sum_{i=1}^{\infty} \epsilon 2^{-i} + \sum_{i=1}^{\infty} i\eta_i \\ &< \epsilon\ (b-a+2) \end{aligned}$$

That is, f is Henstock integrable to F(a,b) on [a,b]. Since the primitive F is absolutely continuous, f is also absolutely Henstock integrable.

To proceed, we need Vitali's covering theorem which we shall state without proof.

Theorem 5.6. If a family of closed intervals covers a set X in the Vitali sense, i.e., for every $x \in X$ and $\eta > 0$ there is a closed interval I in the family such that

$$x \in I \quad \text{and} \quad |I| < \eta,$$

then for every $\epsilon > 0$ there is a finite number of disjoint closed intervals $I_1, I_2, \ldots, I_n$ in the family such that

$$|X| < \sum_{i=1}^{n} |I_i| + \epsilon$$

where $|X|$ denotes the outer measure of X.

We recall that the outer measure of X is defined to be

$$|X| = \inf\{\sum_i |I_i|; \quad \bigcup_i I_i \supset X\}$$

in which I_i denote intervals.

Theorem 5.7. If f is Henstock integrable on [a,b], then its primitive F is differentiable almost everywhere and the derivative $F'(x) = f(x)$ for almost all $x \in [a,b]$.

Proof. Let X be the set of points x at which either $F'(x)$ does not exist or, if it does, is not equal to f(x). We shall prove that X is of measure zero.

From the definition of X we see that for every $x \in X$ there is a $\eta(x) > 0$ such that for every $\delta > 0$ either there is a point u with $0 < x - u < \delta$ and

$$|F(x) - F(u) - f(x)(x-u)| > \eta(x)|x-u|$$

or there is a point v with $0 < v - x < \delta$ and

$$|F(v) - F(x) - f(x)(v-x)| > \eta(x)|v-x|.$$

Fix n and let X_n denote the subset of X for which $\eta(x) \geq 1/n$. Then the above family of closed intervals [u,x] and [x,v] covers X_n in the Vitali sense. Applying the Vitali covering theorem, given $\epsilon > 0$ we can find $[u_k, v_k]$ for $k = 1,2,\ldots, m$ with $u_k = x_k$ or $v_k = x_k$ such that

$$|X_n| < \sum_{k=1}^{m} |v_k - u_k| + \epsilon.$$

Since f is Henstock integrable on [a,b], there is a $\delta(\xi) > 0$ such that for any δ-fine division $D = \{[u,v];\xi\}$ we have

$$\sum|F(v) - F(u) - f(\xi)(v-u)| < \epsilon.$$

When forming the above family of closed intervals, we may assume $\delta \le \delta(x)$. Then we have

$$|X_n| \le \sum_{k=1}^{m} \{|F(v_k) - F(u_k) - f(x_k)(v_k - u_k)|/\eta(x_k)\} + \epsilon$$

$$< \epsilon n + \epsilon$$

Since ϵ is arbitrary, the outer measure of X_n is 0 and so is X.

In conclusion, in order that f should be absolutely Henstock integrable on [a,b] it is necessary and sufficient that its primitive F be absolutely continuous on [a,b] and $F'(x) = f(x)$ almost everywhere. This is the descriptive definition of the Lebesgue integral. Hence we have proved that the Lebesgue and absolute Henstock integrals are equivalent. If we use this as a definition Theorem 5.5 shows that the integral is uniquely determined.

As an application of Theorem 5.7, we can show that if f is Henstock integrable on [a,b] with the primitive F and if $f(x) > 0$ almost everywhere, then the integral $F(a,b) > 0$. If not, then $F(u,v) = 0$ for all u,v. Then by Theorem 5.7 we have $f(x) = F'(x) = 0$ almost everywhere which is a contradiction.

Theorem 5.8. If F is of bounded variation on [a,b] then the derivative $F'(x)$ exists almost everywhere and is absolutely Henstock integrable on [a,b]. Furthermore,

$$\int_a^b |F'(x)|dx \le V(F;[a,b])$$

where V denotes the total variation of F on [a,b].

Proof. Write $V(x) = V(F; [a,x])$ and put $F_1 = (V + F)/2$ and $F_2 = (V-F)/2$. Then both F_1 and F_2 are monotone increasing. Lebesgue's theorem states that every monotone increasing function is differentiable almost everywhere. Therefore both $F_1'(x)$ and $F_2'(x)$ exist almost everywhere. By Fatou's lemma (Corollary 4.4) we obtain

$$\int_a^b F_i'(x)dx \le F_i(b) - F_i(a), \quad i = 1,2.$$

Note that $F = F_1 - F_2$. Consequently, we have

$$\int_a^b |F'(x)|dx \le \int_a^b F_1'(x)dx + \int_a^b F_2'(x)dx$$

$$\le V(F;[a,b]).$$

The proof is complete.

Theorem 5.9. If f is absolutely Henstock integrable on [a,b], then for every $\epsilon > 0$ there is a step function φ such that

$$\int_a^b |f(x)-\varphi(x)|dx < \epsilon.$$

Proof. Suppose f is bounded on [a,b]. Define F_n in terms of the primitive F of f as follows:

$$F_n(x) = F(x) \quad \text{when} \quad x = a + i(b-a)2^{-n},\ i = 0, 1, 2, \ldots, 2^n,$$

and linearly elsewhere in [a,b]. Let $f_n(x) = F_n'(x)$ almost everywhere. Note that each F_n is continuous and piecewise linear. Therefore f_n is a step function taking values of the form $(F(y) - F(x))/(y-x)$. Since F is differentiable almost everywhere, the sequence of step functions $f_n(x)$ converges to f(x) almost everywhere. We may assume that the sequence $\{f_n\}$ is uniformly bounded. Hence by the dominated convergence theorem we have

$$\int_a^b |f_n - f| \longrightarrow 0 \text{ as } n \longrightarrow \infty.$$

In general, given $\epsilon > 0$ we can find a truncated function f^N (Lemma 5.4) and consequently a step function φ such that

$$\int_a^b |f-\varphi| \le \int_a^b |f-f^N| + \int_a^b |f^N-\varphi| < 2\epsilon.$$

The proof is complete.

We shall return to the subject of absolute integrability in Section 16.

We mention here another application of Theorem 5.7.

Theorem 5.10. If f is Henstock integrable on [a,b], then f is almost eveywhere the limit of a sequence of step functions.

Proof. Let F be the primitive of f. Define F_n in terms of F as in the proof of Theorem 5.9. Again, let $f_n(x) = F_n'(x)$ almost everywhere. Apply Theorem 5.7 and we obtain that f is almost everywhere the limit of a sequence of step functions $\{f_n\}$.

We end this chapter with a definition of measurable functions and the relation with measurable sets. A function is said to be measurable if it is almost everywhere the limit of a sequence of step functions. What we have proved above is that every Henstock integrable function is necessarily measurable.

Theorem 5.11. If f is measurable and $g(x) \le f(x) \le h(x)$ almost everywhere in [a,b] where g and h are Henstock integrable on [a,b], then f is Henstock integrable on [a,b].

Proof. We may assume that $g(x) = 0$ for all x. If not, consider $f - g$ in place of f. Let $\{\varphi_n\}$ be a sequence of step functions converging to f pointwise almost everywhere. Consider $f_n = \min(h, \varphi_n)$. Then $f_n(x) \longrightarrow f(x)$ almost everywhere as $n \to \infty$ and the sequence $\{f_n\}$ is dominated by h. Hence the integrability of f follows from the dominated convergence theorem (Theorem 4.3).

Let $X \subset [a,b]$. The set X is said to be measurable if the characteristic function χ_X is measurable. The measure of X is defined to be

$$|X| = \int_a^b \chi_X(x)\,dx.$$

The set of measure zero here coincides with that of Example 2.5. Also, the measure of X coincides with the outer measure of X if X is measurable. A simple function is a function which takes constant values on a finite number of pairwise disjoint measurable sets. A step function is a simple function but not conversely. Obviously, a simple function is measurable.

Theorem 5.12. A function f is measurable on [a,b] if and only if for every real number c the set

$$X(f<c) = \{x \in [a,b];\ f(x) < c\}$$

is a measurable set in [a,b].

Proof. Suppose f is measurable. Then there is a sequence of step functions $\{\varphi_n\}$ such that $\varphi_n(x) \to f(x)$ almost everywhere as $n\to\infty$, i.e., everywhere except in S with $|S| = 0$. Let

$$X = X(f<c), \quad Y = \bigcup_{k=1}^{\infty} \bigcup_{n=1}^{\infty} \bigcap_{i=n}^{\infty} X(\varphi_i < c - \frac{1}{k}).$$

Then it is easy to verify that $X-S = Y-S$. Applying the monoton convergence theorem repeatedly, we see that the characteristic function of Y is Henstock integrable and therefore measurable on $[a,b]$. Hence X is measurable since it differs from Y only on a set of measure zero.

Conversely, suppose $X(f<c)$ is measurable for all c. We may assume that f is non-negative. Write

$$X_{ni} = \{x \in [a,b];\ (i-1)2^{-n} \leq f(x) < i2^{-n}\}, \quad i = 1,2,3,\ldots,n2^n,$$

$$X_n = \{x \in [a,b];\ f(x) \geq n\}.$$

Then they are pairwise disjoint measurable sets in $[a,b]$. Define a simple function f_n as follows:

$$f_n(x) = (i-1)2^{-n} \text{ when } x \in X_{ni},\ i = 1,2,3,\ldots,n2^n,$$

$$= n \qquad \text{when } x \in X_n.$$

Obviously, $f_n(x) \to f(x)$ pointwise as $n\to\infty$.

In view of Theorem 5.9, for each n there is a step function φ_n such that

$$\int_a^b |f_n-\varphi_n| < 4^{-n}.$$

Let S_n denote the set of all x for which $|f_n(x)-\varphi_n(x)| \geq 2^{-n}$. Then

$$2^{-n}|S_n| \leq \int_{S_n} |f_n-\varphi_n| < 4^{-n},$$

and consequently $|S_n| < 2^{-n}$. Obviously,

$$|f_n(x) - \varphi_n(x)| < 2^{-n} \quad \text{for } x \in [a,b]-S_n,\ n = 1,2,\ldots .$$

Therefore for every m and for every $x \in [a,b] - \cup_{n=m}^{\infty} S_n$ the sequence $f_n(x)-\varphi_n(x)$ converges to 0 as $n\to\infty$. Let S be the set of all x for which $f_n(x)-\varphi_n(x)$ does not converge or if it does but not to 0. It follows that for every m

$$S \subset \bigcup_{n=m}^{\infty} S_n, \qquad |\bigcup_{n=m}^{\infty} S_n| < 2^{-m+1}.$$

Consequently, $|S| = 0$. That is, $f_n(x)-\varphi_n(x) \to 0$ almost everywhere in [a,b] as $n\to\infty$. Finally, combining with the fact that $f_n(x)$ converges pointwise to $f(x)$, we have $\varphi_n(x) \to f(x)$ almost everywhere as $n\to\infty$. Hence we have proved that f is measurable.

We may prove other properties involving measurable functions or measurable sets. As we can see here and in the later chapters, the measurability does not play such a major role as in the theory of Lebesgue integration.

CHAPTER 2 THE DENJOY-PERRON INTEGRAL

6. THE DENJOY INTEGRAL

We show in this section that the Denjoy and Henstock integrals are equivalent. Other equivalent definitions will be given in Sections 8, 9, 10 and 17. A series of generalized convergence theorems will also be introduced throughout the chapter, with their relationship discussed in Section 10. First, we recall that absolutely continuous functions have been defined in Defintion 5.1.

Lemma 6.1. A function F is absolutely continuous on [a,b] if and only if for every $\epsilon > 0$ there is a $\eta > 0$ such that for every finite or infinite sequence of non overlapping intervals $\{[a_i,b_i]\}$ satisfying

$$\sum_i |b_i - a_i| < \eta \qquad \text{we have} \qquad \sum_i \omega(F;[a_i,b_i]) < \epsilon$$

where ω denotes the oscillation of F over $[a_i,b_i]$, i.e.,

$$\omega(F;[a_i,b_i]) = \sup\{|F(x) - F(y)|;\ x,y \in [a_i,b_i]\}.$$

The proof is elementary.

Definition 6.2. Let $X \subset [a,b]$. A function F is said to be AC(X) if Definition 5.1 holds with the end points a_i, $b_i \in X$ for all i.

Definition 6.3. Let $X \subset [a,b]$. A function F is said to be AC*(X) if the condition in Lemma 6.1 holds with the end points a_i, $b_i \in X$ for all i.

We remark that AC(X) and AC*(X) no longer coincide when X ceases to be an interval.

Lemma 6.4. Let X be a closed set in [a,b] and (a,b) − X the union of (c_k,d_k) for k = 1, 2, ... Suppose F is continuous on [a,b]. Then the following conditions are equivalent:

(i) F is AC*(X),

(ii) F is AC(X) and

$$\sum_{k=1}^{\infty} \omega(F;[c_k,d_k]) < +\infty,$$

(iii) Definition 5.1 holds with a_i or b_i belonging to X for every i.

Proof. It is easy to see that (i) implies (ii). To prove (ii) implying (iii), for every $\epsilon > 0$ we choose N such that

$$\sum_{k=N+1}^{\infty} \omega(F;[c_k,d_k] < \epsilon.$$

Since F is AC(X), there is a $\eta > 0$ such that for every finite or infinite sequence of non-overlapping intervals $\{[a_i,b_i]\}$ with $a_i,b_i \in X$ and satisfying

$$\sum_i |b_i - a_i| < \eta \qquad \text{we have} \quad \sum_i |F(b_i) - F(a_i)| < \epsilon.$$

Since F is continuous, we may choose the same $\eta > 0$ such that whenever ξ is one of the points c_k and d_k, $k = 1, 2, .., N$, and

$$|\xi - t| < \eta \quad \text{we have} \quad |F(\xi) - F(t)| < \epsilon/N.$$

Now take a finite or infinite sequence of non-overlapping intervals $\{[a_i,b_i]\}$ with the total length less than η with a_i or b_i belonging to X for every i. For example, if $a_i \in X$, then b_i either belongs to X or lies in (c_k,d_k) for some k. Considering various cases and using the above three inequalities, we obtain

$$\sum_i |F(b_i) - F(a_i)| < 4\epsilon.$$

Hence we have proved (iii).

We observe that since F is continuous, we have

$$\omega(F;[a_i,b_i]) = |F(\beta_i) - F(\alpha_i)|$$

for some α_i and $\beta_i \in [a_i,b_i]$. Therefore (i) follows from (iii) in view of the following inequality

$$\omega(F_i[a_i,b_i]) \le |F(\alpha_i) - F(a_i)| + |F(\beta_i) - F(a_i)|$$

in which a_i on the right side of the above inequality may be replaced by b_i.

We remark that under the continuity condition, to proceed from AC(X) to AC*(X) we change either from the difference $F(y) - F(x)$ to the oscillation $\omega(F;[x,y])$ or from the two endpoints of the intervals belonging to X to at least one endpoint belonging to X.

As we shall see, the Denjoy integral is a countable extension of the Lebesgue integral (Section 5). First, we consider a countable extension of AC*(X).

Definition 6.5. A function F is said to be ACG* on X if X is the union of a sequence of sets $\{X_i\}$ such that on each X_i the function F is AC*(X_i).

Example 6.6. Let F be continuous on [a,b] and differentiable nearly everywhere in [a,b], i.e., everywhere except perhaps for a countable number of points in [a,b]. We shall show that F is ACG* on [a,b].

We observe that if F is continuous at c and AC*(X) then F is also AC*($X \cup \{c\}$). For simplicity, we may assume that F is differentiable everywhere. Let X_{ni} denote the set of points $x \in [a+(i-1)/n, a+i/n]$ and $x \in [a,b]$ such that

$$|F(t) - F(x)| \le n(t - x) \quad \text{whenever} \quad 0 \le t - x \le 1/n.$$

Obviously, [a,b] is the union of X_{ni} for i = 1, 2, ...; n = 1, 2, ... Take any sequence of intervals I_k with the endpoints in X_{ni}. Then $\omega(F;I_k) \le 2n|I_k|$ and for every $\epsilon > 0$ choose $\eta > 0$ with $2n\,\eta < \epsilon$ so that

$$\sum_k |I_k| < \eta \quad \text{implies} \quad \sum_k \omega(F;I_k) \le 2n \sum_k |I_k| < \epsilon.$$

That is, F is AC*(X_{ni}). Consequently, F is ACG* on [a,b].

In particular, the function $F(x) = x^2 \sin x^{-2}$ when $x \ne 0$ and F(0) = 0 as given in Example 1.3 is ACG* on [0,1]. However not every continuous function is ACG^*, for example, the function F given in Example 11.1 is continuous on [0,1] but not ACG^* there.

Theorem 6.7. Let $X \subset [a,b]$. If F is continuous on [a,b] and AC*(X), then F is AC*($\bar{X}$) where $\bar{X}$ is the closure of X.

The proof is again elementary.

In what follows, we shall consider only continuous functions that are ACG*. Hence we may always assume that [a,b] is the union of a sequence of closed sets X_i, i = 1, 2, ..., on each of which F is AC*(X_i). In fact, this is the case in the following definition of the Denjoy integral.

Definition 6.8. A function f is said to be Denjoy integrable on [a,b] if there is a function F which is continuous and ACG* on [a,b] such that its derivative $F'(x) = f(x)$ almost everywhere in [a,b].

Example 6.9. Let f be nearly everywhere the derivative of a continuous function on [a,b]. Then by Example 6.6 the function f is Denjoy integrable. In particular, every Newton integrable function is Denjoy integrable.

Example 6.10. Every absolutely Henstock integrable function on [a,b] is Denjoy integrable.

The uniqueness of the Denjoy integral follows from the following theorem.

Theorem 6.11. If F is continuous and ACG* on [a,b] and if its derivative $F'(x) \geq 0$ almost everywhere in [a,b] then F is monotone increasing.

Proof. Let $F'(x) \geq 0$ for $x \in [a,b] - S$ where S is of measure zero. Suppose $x \in [a,b] - S$. Then for every $\epsilon > 0$ there exists a $\delta(x) > 0$ such that whenever $x \in [u,v] \subset (x - \delta(x), x + \delta(x))$ we have

$$F(v) - F(u) > - \epsilon(v - u).$$

Next, suppose $x \in S$. Since F is continuous and ACG* on [a,b], then by Lemma 6.4 (iii) and Theorem 6.7 the interval [a,b] is the union of a sequence of closed sets $\{X_i\}$ such that F is $AC*(X_i)$ for each i, i.e., for every $\epsilon > 0$ there is a $\eta_i > 0$ such that for every sequence of non-overlapping intervals $\{I_k\}$ with at least one endpoint of I_k belonging to X_i and satisfying

$$\sum_k |I_k| < \eta_i \qquad \text{we have} \qquad \sum_k |F(I_k)| < \epsilon 2^{-i}.$$

Here F(I) denotes F(v)−F(u) where I = [u,v]. Write $S_i = S \cap Y_i$ where $Y_1 = X_1$ and $Y_i = X_i - (X_1 \cup X_2 \cup \ldots \cup X_{i-1})$ for i = 2, 3, ... Choose open intervals I_{ij}, j = 1, 2, ..., such that

$$\sum_j |I_{ij}| < \eta_i \qquad \text{and} \qquad \bigcup_j I_{ij} \supset S_i.$$

Now for $x \in S$ put $(x - \delta(x), x + \delta(x)) \subset I_{ij}$ for some j. Hence we have defined a positive function $\delta(x)$.

Take a δ-fine division $D = \{[u,v];\xi\}$. Applying the inequalities above, we have

$$F(b) - F(a) = \sum\{F(v) - F(u)\} > -\epsilon(b-a)-2\epsilon.$$

Since ϵ is arbitrary, we have $F(b) - F(a) \geq 0$. The same holds for any subinterval of $[a,b]$. Hence F is monotone increasing on $[a,b]$.

Theorem 6.12. If f is Denjoy integrable on $[a,b]$, then it is Henstock integrable there.

Proof. The proof is a combination of Theorems 5.5 and 6.11. Let F be the primitive of f and $F'(x) = f(x)$ for $x \in [a,b] - S$ where S is of measure zero. For $\xi \in [a,b] - S$, given $\epsilon > 0$ there is a $\delta(\xi) > 0$ such that whenever $\xi \in [u,v] \subset (\xi - \delta(\xi), \xi + \delta(\xi))$ we have

$$|F(u,v) - f(\xi)(v - u)| \leq \epsilon\ |v - u|.$$

Since F is continuous and ACG*, there is a sequence of closed sets $\{X_i\}$ such that $\cup_i X_i = [a,b]$ and F is $AC*(X_i)$ for each i. Let $Y_1 = X_1$, $Y_i = X_i - (X_1 \cup X_2 \cup \ldots X_{i-1})$ for $i = 2, 3, \ldots$ and S_{ij} denote the set of points $x \in S \cap Y_i$ such that $j - 1 \leq |f(x)| < j$. Obviously, S_{ij}, $i, j = 1, 2, \ldots$, are pairwise disjoint and their union is the set S. Since F is also $AC*(S_{ij})$, there is a $\eta_{ij} < \epsilon 2^{-i-j}j^{-1}$ such that for any sequence of non-overlapping intervals $\{I_k\}$ with at least one endpoint of I_k belonging to S_{ij} and satisfying

$$\sum_k |I_k| < \eta_{ij} \quad \text{we have} \quad \sum_k |F(I_k)| < \epsilon 2^{-i-j}.$$

Again, $F(I)$ denotes $F(v)-F(u)$ where $I = [u,v]$. Choose G_{ij} to be the union of a sequence of open intervals such that

$$|G_{ij}| < \eta_{ij} \quad \text{and} \quad G_{ij} \supset S_{ij}$$

where $|G_{ij}|$ denotes the total length of G_{ij}. Now for $\xi \in S_{ij}$, $i, j = 1, 2, \ldots$, put $(\xi - \delta(\xi), \xi + \delta(\xi)) \subset G_{ij}$. Hence we have defined a positive function $\delta(\xi)$.

Take any δ-fine division $D = \{[u,v]; \xi\}$. Split the sum $\sum$ over D into two partial sums $\sum_1$ and $\sum_2$ in which $\xi \notin S$ and $\xi \in S$ respectively and we obtain

$$|\sum f(\xi)(v - u) - F(a,b)| \leq \sum_1 |f(\xi)(v - u) - F(u,v)|$$
$$+ \sum_2 |F(u,v)| + \sum_2 |f(\xi)(v - u)|$$

$$< \epsilon(b - a) + \sum_{i,j} \epsilon 2^{-i-j} + \sum_{i,j} j\eta_{ij}$$

$$< \epsilon(b - a) + 2\epsilon.$$

That is, f is Henstock integrable to F(a,b) on [a,b].

Theorem 6.13. If f is Henstock integrable on [a,b] then it is Denjoy integrable there.

We shall verify the result in a series of definitions and lemmas as follows.

Definition 6.14. **Let X ⊂** [a,b]. A function F is said to be VB*(X) if

$$\sup \sum_k \omega(F;[a_k,b_k]) < +\infty$$

where ω denotes the oscillation of F on $[a_k,b_k]$ and the supremum is taken over all finite or infinite sequence of non-overlapping intervals $\{[a_k,b_k]\}$ with a_k, $b_k \in X$ for all k. A function F is said to be VBG* on X if X is the union of a sequence of sets $\{X_i\}$ such that F is VB*(X_i) for each i.

Lemma 6.15. Let X ⊂ [a,b] and a,b ∈ X. Then a function F is VB*(X) if and only if Definition 6.14 holds with a_k, $b_k \in X$ replaced by a_k or $b_k \in X$, i.e. at least one of a_k and b_k belonging to X.

Proof. The sufficiency is trivial. Suppose F is VB*(X). Take any sequence of non-overlapping intervals $\{[a_k,b_k]\}$ with a_k or b_k $\in X$ for all k. Sort out the intervals into two classes, one with the left endpoint a_k in X and another with the right endpoint b_k in X. Then it follows that

$$\sum_k \omega(F;[a_k,b_k]) \le 2M$$

where M stands for the supremum as given in Definition 6.14. Hence the condition is necessary.

As a consequence of Lemma 6.15, we have the following

Lemma 6.16. Let X ⊂ [a,b] and a,b ∈ X. If F is VB*(X) then it is VB*($\overline{X}$) where $\overline{X}$ is the closure of X.

We remark that we did not assume the continuity of F in the above two lemmas whereas we did in the case of AC*(X). Also, if we require only one endpoint a_k or b_k to belong to X in Definition 6.14, then we may replace the oscillation by the difference $|F(b_k) - F(a_k)|$.

Lemma 6.17. If f is Henstock integrable on [a,b] with the primitive F then F is VBG* on [a,b].

Proof. Since f is Henstock integrable on [a,b], for every $\epsilon > 0$ there is a $\delta(\xi) > 0$ such that for any δ-fine division $D = \{[u,v];\xi\}$ we have

$$\sum |F(u,v) - f(\xi)(v - u)| < \epsilon.$$

For convenience, we may put $\epsilon = 1$ and $\delta(\xi) \le 1$. Let X_{ni} denote the set of all points $x \in [a+(i-1)/n, a+i/n]$ and $x \in [a,b]$ such that

$$|f(x)| \le n \quad \text{and} \quad 1/n < \delta(\xi) \le 1/(n-1).$$

Obviously, the union of X_{ni}, $i = 1, 2, \ldots, n = 2, 3, \ldots$, is [a,b]. Take any finite sequence of non-overlapping intervals $\{[a_k,b_k]\}$ with $a_k, b_k \in X_{ni}$ for all k. Then the intervals form a δ-fine partial division of [a,b]. Writing $a_k \le u_k \le v_k \le b_k$, we have

$$\sum_k |F(u_k,v_k)| \le \sum_k |F(a_k,b_k)| + \sum_k |F(a_k,u_k)| + \sum_k |F(v_k,b_k)|$$

$$< 3\epsilon + 2\sum_k |f(a_k)(b_k-a_k)| + \sum_k |f(b_k)(b_k-v_k)|$$

$$< 3 + 3n(b-a)$$

That is, F is VB*(X_{ni}). Consequently, F is VBG* on [a,b].

Lemma 6.18. Let f be Henstock integrable on [a,b] with the primitive F. If F is VB*(X) and X is closed, then f_X is absolutely Henstock integrable on [a,b] where $f_X(x) = f(x)$ when $x \in X$ and 0 otherwise.

Proof. Since f is Henstock integrable on [a,b], for every $\epsilon > 0$ there is a $\delta(\xi) > 0$ such that for any δ-fine division $D = \{[u,v];\xi\}$ we have

$$\sum |F(u,v) - f(\xi)(v - u)| < \epsilon.$$

Let (a,b) − X be the union of (c_k,d_k), k = 1, 2, ..., and write

$$A = F(a,b) - \sum_{k=1}^{\infty} F(c_k,d_k).$$

The above series converges because F is VB*(X). Choose a sufficiently large N such that

$$\sum_{k=N+1}^{\infty} \omega(F;[c_k,d_k]) < \epsilon .$$

Choose again a $\eta > 0$ such that when x is one of the points c_k and d_k for $k = 1, 2, \ldots, N$,

$$|t - x| < \eta \quad \text{implies} \quad |F(t) - F(x)| < \epsilon/N.$$

Now modify $\delta(\xi)$ as follows. Assume $(\xi-\delta(\xi), \xi + \delta(\xi)) \subset (c_k,d_k)$ if $\xi \in (c_k,d_k)$ for some k. Define an open set $G \supset X$ by

$$G = X\cup(\bigcup_{k=N+1}^{\infty} (c_k,d_k)) \cup (\bigcup_{k=1}^{N}(c_k,c_k+\eta)) \cup (\bigcup_{k=1}^{N}(d_k-\eta,d_k)).$$

Assume $(\xi-\delta(\xi), \xi + \delta(\xi)) \subset G$ if $\xi \in X$. Hence we have modified $\delta(\xi)$. Take any δ-fine division $D = \{[u,v];\xi\}$. Split the sum $\sum$ over D into two partial sums $\sum_1$ and $\sum_2$ in which $\xi \in X$ and $\xi \notin X$ respectively. Then it follows that

$$\begin{aligned}|\sum f_X(\xi)(v-u) - A| &\le | \sum_1 f(\xi)(v-u) - \sum_1 F(u,v)| + |\sum_1 F(u,v) - A| \\ &< \epsilon + |\sum_2 F(u,v) - \sum_{k=1}^{\infty} F(c_k,d_k)|.\end{aligned}$$

Note that both endpoints of $[u,v]$ in $\sum_2$ do not lie in X. In view of the way we modified $\delta(\xi)$, the last term above is less than 3ϵ. Therefore f is Henstock integrable on [a,b]. The absolute integrability follows from Lemma 4.2a.

Lemma 6.19. If f is Henstock integrable on [a,b] with the primitive F, then F is ACG* on [a,b].

Proof. First, we have shown that F is VBG* (Lemma 6.17). We may assume that each X_i in the definition of VBG* is closed (Lemma 6.16). In view of Corollary 3.8 and Lemma 6.4(ii), it suffices to show that F is $AC(X_i)$ for each i. If so, then F is ACG*.

Write $X = X_i$ and let F_X be the primitive of f_X as given in Lemma 6.18. Using the notation of Lemma 6.18, we have

$$F(x) = F_X(x) + G(x) \text{ and } G(x) = \sum_{k=1}^{\infty} F([a,x] \cap [c_k,d_k]) \text{ for } x \in [a,b].$$

Since f_X is absolutely Henstock integrable, F_X is absolutely continuous. To show that G is AC(X), choose N as in the proof of Lemma 6.18 and let

$$0 < \eta < \min\{|d_k - c_k|;\ k = 1, 2, \ldots, N\}.$$

Then for any sequence of non-overlapping intervals $\{I_k\}$ with both endpoints of I_k belonging to X and satisfying

$$\sum_k |I_k| < \eta \qquad \text{we have} \quad \sum_k \omega(G; I_k) < \epsilon.$$

That is, G is AC(X). Consequently, F is AC(X). The proof is complete.

Proof of Theorem 6.13. Let F be the primitive of f. By Corollary 3.8 and Lemma 6.19, F is continuous and ACG^*. Again, by Theorem 5.7, the derivative $F'(x) = f(x)$ almost everywhere. Hence by definition f is Denjoy integrable on [a,b].

Finally, we have completed the proof that f is Henstock integrable if and only if it is Denjoy integrable. Furthermore, they have the same integral. It is possible to develop the theory of integration using the definition of the Denjoy integral though the proofs would normally be less elementary.

Here we give a characterization of the property ACG^*.

Definition 6.20. A function F is said to satisfy the condition (N) if $|F(E)| = 0$ whenever $|E| = 0$, where $F(E) = \{F(x);\ x \in E\}$.

We remark that in the proof of Theorem 6.12 it is the condition (N) that is used and not the full property of ACG^*.

Theorem 6.21. Let F be a continuous function on [a,b]. Then F is ACG^* if and only if it is VBG^* and satisfies the condition (N).

Proof. Suppose F is VBG^* and satisfies the condition (N). Then F is $VB^*(X_i)$ for each i where $\cup_i X_i = [a,b]$ and each X_i is closed. Fix i and write $X = X_i$. We may assume $a,b \in X$. Put $G(x) = F(x)$ when $x \in X$ and linear elsewhere. Then G is of bounded variation on [a,b]. It follows from Theorem 5.8 that $G'(x) = F'(x)$ exists almost everywhere in X. Consequently, $F'(x)$ exists almost everywhere in [a,b].

Now put $f(x) = F'(x)$ when $F'(x)$ exists. Then using the proof of Theorem 6.12 and the remark after Definition 6.20 we obtain that f is Henstock integrable on [a,b] with the primitive F. Hence by Lemma 6.19

the function F is ACG^*. The converse is easy.

Theorem 6.22. In order that f be Henstock integrable on [a,b] it is necessary and sufficient that there exists a continuous function F satisfying the condition (N) such that the derivative $F'(x)$ exists and equals $f(x)$ almost everywhere in [a,b].

This is a re-statement of Theorems 6.12 and 6.13. However we may also prove directly as follows that if f is Henstock integrable on [a,b] then its primitive F satisfies the condition (N). Suppose $|E| = 0$. We may assume that $f(x) = 0$ for $x \in E$. Since f is Henstock integrable on [a,b] for every $\epsilon > 0$ there is a $\delta(\xi) > 0$ such that for any δ-fine division $D = \{[u,v];\xi\}$ we have

$$\sum|F(u,v)-f(\xi)(v-u)| < \epsilon$$

where $\sum$ sums over D and $F(u,v)$ denotes $F(v)-F(u)$. If $\sum_1$ is a partial sum of the above $\sum$ in which $\xi \in E$ then

$$\sum_1|F(u,v)| < \epsilon.$$

For each $\xi \in E$ there is a shrinking family of intervals [u,v] such that $\xi \in [u,v] \subset (\xi-\delta(\xi),\ \xi+\delta(\xi))$. Then the collection M of such intervals forms a Vitali covering of E. Since F is continuous, $F([u,v]) = [F(u^*),F(v^*)]$ where $F(u^*)$ and $F(v^*)$ are respectively the infimum and the supremum of $F(x)$ for $x \in [u,v]$. In view of the continuity of F again, the family of all intervals $F([u,v])$ for $[u,v] \in M$ forms a Vitali covering of F(E). Thus by the Vitali covering theorem (Theorem 5.6), there is a finite sum $\sum_2$ such that

$$\begin{aligned}|F(E)| &< \sum_2|F([u,v])| + \epsilon \\ &\le \sum_2|F(v^*)-F(\xi)| + \sum_2|F(\xi)-F(u^*)| + \epsilon \\ &< 3\epsilon.\end{aligned}$$

Since ϵ is arbitrary, $|F(E)| = 0$ and the condition (N) is satisfied.

7. THE CONTROLLED CONVERGENCE THEOREM

We shall prove in this section a generalized convergence theorem, known as the controlled convergence theorem. We shall give three different proofs; one in this section by the Denjoy integral, another in Section 8 by the standard category argument, and finally a direct proof in Section 9. Other convergence theorems, namely, the generalized dominated convergence theorem (Theorem 8.12) and the generalized mean convergence theorem (Theorem 9.7) will appear later. Their relationship will be discussed in Section 10. First, we prove a special case, namely, Vitali's convergence theorem. A family of functions $\{F_n\}$ is said to be uniformly absolutely continuous if F_n is absolutely continuous but uniformly in n, i.e., $\eta > 0$ in Definition 5.1 with F replaced by F_n is independent of n.

Theorem 7.1. If the following conditions are satisfied:

(i) $f_n(x) \to f(x)$ almost everywhere in $[a,b]$ as $n \to \infty$ where each f_n is Henstock integrable on $[a,b]$;

(ii) the primitives F_n of f_n are uniformly absolutely continuous,

then f is Henstock integrable on $[a,b]$ and

$$\int_a^b f_n \to \int_a^b f \qquad \text{as } n \to \infty.$$

To proceed, we need the following lemmas, the first of which is a special version of Egoroff's theorem.

Lemma 7.2. If condition (i) in Theorem 7.1 holds, then for every $\eta > 0$ there is an open set G with $|G| < \eta$ such that $f_n(x)$ converges uniformly to $f(x)$ on $[a,b] - G$.

For a proof of Egoroff's theorem, see any standard textbook on real analysis. We shall also give a proof of Theorem 7.1 without reference to Egoroff's theorem.

Lemma 7.3. If the conditions in Theorem 7.1 hold, then for every $\epsilon > 0$ there is an integer N such that for every partial division $D = \{[u,v]\}$ of $[a,b]$ we have

$$|\sum\{F_n(u,v) - F_m(u,v)\}| < \epsilon \qquad \text{whenever } n, m > N.$$

Proof. Since F_n are uniformly absolutely continuous, for every $\epsilon > 0$ there is a $\eta > 0$ (independent of n) such that for any sequence of non-overlapping intervals $\{[a_i,b_i]\}$ satisfying

$$\sum_i |b_i - a_i| < \eta \quad \text{we have} \quad \sum_i |F_n(a_i,b_i)| < \epsilon.$$

In view of Lemma 7.2, there are an integer N and an open set G with

$|G| < \eta$ such that

$$|f_n(x) - f_m(x)| < \epsilon \qquad \text{for } x \in [a,b] - G \text{ and all } n,\ m \geq N.$$

Further, for $n, m \geq N$ we choose a common $\delta(\xi) > 0$, depending on n and m, such that for any δ-fine division $D = \{[u,v];\ \xi\}$ we have

$$\sum|F_i(u,v) - f_i(\xi)(v - u)| < \epsilon \qquad \text{for } i = n,\ m.$$

Modify the above $\delta(\xi)$ so that $(\xi - \delta(\xi),\ \xi + \delta(\xi)) \subset G$ when $\xi \in G$,

Now consider a partial division D of $[a,b]$ and to avoid ambiguity let $[x,y]$ be a typical interval in D. Next, consider a δ-fine division $D_1 = \{[u,v];\xi\}$ of $[x,y]$ and $\sum_1$ over D_1. Write $\sum_1 = \sum_2 + \sum_3$ where $\sum_2$ denotes the partial sum of $\sum_1$ for which $\xi \in [a,b] - G$ and $\sum_3$ otherwise. Then we have

$$\begin{aligned} |F_n(x,y) - F_m(x,y)| &= |\textstyle\sum_1\{F_n(u,v) - F_m(u,v)\}| \\ &\leq \textstyle\sum_2|F_n(u,v) - f_n(\xi)(v-u)| + \sum_2|f_m(\xi)(v-u) - F_m(u,v)| \\ &\quad + \textstyle\sum_2|f_n(\xi) - f_m(\xi)|(v-u) + \sum_3|F_n(u,v)| + \sum_3|F_m(u,v)|. \end{aligned}$$

Making use of all the above inequalities and writing $\sum$ over D, we obtain for $n, m \geq N$

$$|\sum\{F_n(x,y) - F_m(x,y)\}| < 2\epsilon + \epsilon(b-a) + 2\epsilon.$$

That is, the required condition holds.

Proof of Theorem 7.1. It follows from Lemma 7.3 that

$$F(x,y) = \lim_{n\to\infty} F_n(x,y) \qquad \text{exists}$$

for any subinterval $[x,y]$ of $[a,b]$. Applying Lemma 7.3 again, we can find a subsequence $F_{n(j)}$ of F_n such that

$$|\sum\{F_{n(j)}(u,v) - F(u,v)\}| < \epsilon 2^{-j} \qquad \text{for } j = 1,2,\ldots,$$

where $\sum$ sums over any partial division $D = \{[u,v]\}$ of $[a,b]$.

Following the proof of the monotone convergence theorem (Theorem 4.1) with f_n replaced by $f_{n(j)}$ and using the notation there, we are to prove

$$|\sum\{F_{m(\epsilon,\xi)}(u,v) - F(u,v)\}| < \epsilon$$

where $\sum$ sums over any δ-fine division $D = \{[u,v];\xi\}$. Indeed, this follows easily from what we have proved above. Hence f is Henstock integrable to $F(a,b)$ on $[a,b]$ and $F_n(a,b)$ converges to $F(a,b)$ as $n \to \infty$.

We remark that if the sequence $\{f_n\}$ is dominated by a Henstock integrable function f, i.e. $|f_n(x)| \le g(x)$ almost everywhere for all n and g is Henstock integrable, then the primitives F_n of f_n are uniformly absolutely continuous. Therefore the dominated convergence theorem (Theorem 4.3) may now follow from Theorem 7.1 as a corollary, by writing

$$0 \le f_n(x) - g(x) \le h(x) - g(x)$$

and considering $f_n - g$ in place of f_n.

An alternative proof of Theorem 7.1. By Lemma 4.2a, the function $\min\{f_n; i \le n \le j\}$ is Henstock integrable on $[a,b]$. The rest of the proof is identical to that of Theorem 4.3.

Note that the alternative proof did not use Egoroff's theorem.

Definition 7.4. A sequence of functions $\{f_n\}$ is said to be control-convergent to f on $[a,b]$ if the following conditions are satisfied:

(i) $f_n(x) \to f(x)$ almost everywhere in $[a,b]$ as $n \to \infty$ where each f_n is Henstock integrable on $[a,b]$;

(ii) the primitives F_n of f_n are ACG* uniformly in n, i.e., $[a,b]$ is the union of a sequence of closed sets X_i such that on each X_i the functions F_n are $AC^*(X_i)$ uniformly in n, in other words, $\eta > 0$ in the definition of $AC^*(X_i)$ is independent of n;

(iii) the primitives F_n converge uniformly on $[a,b]$.

Example 7.5. Let $F(0) = 0$ and

$$F(x) = x^2 \sin x^{-2} \quad \text{when } x \ne 0.$$

Let $f(x) = F'(x)$ for $x \in [0,1]$, also $f_n(x) = f(x)$ when $1/n \le x \le 1$ and 0 otherwise, $n = 1, 2, \ldots$. Then f_n is control-convergent to f on $[0,1]$ though the sequence is not dominated by any Henstock integrable function on the left or on the right.

Now we state and give the first proof of the controlled convergence theorem.

Theorem 7.6. If a sequence of functions $\{f_n\}$ is control-convergent to f on $[a,b]$, then f is Henstock integrable on $[a,b]$ and we have

$$\int_a^b f_n \longrightarrow \int_a^b f \qquad \text{as } n \to \infty.$$

Proof. In view of Definition 7.4(iii), $F(x)$ exists as the limit of $F_n(x)$ and is continuous. It follows from Definition 7.4(ii) that F is ACG*. It remains to show that $F'(x) = f(x)$ almost everywhere.

Suppose F_n are AC*(X) uniformly in n with X closed. For convenience, we assume $a,b \in X$. If we can show that $F'(x) = f(x)$ almost everywhere in X, then the proof is complete in view of (ii). To do so, we put $G_n(x) = F_n(x)$ when $x \in X$ and linearly elsewhere in $[a,b]$, or more precisely, on the intervals (u,v) of $(a,b) - X$ with $u, v \in X$ we take G_n linear from $G_n(u)$ to $G_n(v)$. Note that $G_n(a) = F_n(a)$ and $G_n(b) = F_n(b)$. Then G_n are uniformly absolutely continuous on $[a,b]$. Further, we put $G(x) = F(x)$ when $x \in X$ and linearly elsewhere in $[a,b]$. Writing $g_n(x) = G_n'(x)$ for almost all x, $g(x) = f(x)$ when $x \in X$ and $G'(x)$ elsewhere, we see that the conditions in Theorem 7.1 are satisfied with f_n and f replaced by g_n and g. Hence we obtain $F'(x) = G'(x) = f(x)$ for almost all x in X. Since the Denjoy and Henstock integrals are equivalent, f is Henstock integrable on $[a,b]$ and the proof is complete.

We remark that in the proof the uniform convergence of F_n is used only to ensure the continuity of F. Hence we have proved the following

Corollary 7.7. Theorem 7.6 holds with condition (iii) in the controlled convergence replaced by: the primitives $F_n(x)$ converge pointwise to a continuous function $F(x)$ at each x in $[a,b]$.

Example 7.8. Let $F_n(x) = \sin 2n\pi x$ when $0 \le x \le 1/n$ and 0 elsewhere and $f_n(x) = F_n'(x)$. Then the sequence $\{f_n\}$ satisfies the conditions in Corollary 7.7 but not those in Theorem 7.6 with $f(x) = 0$ for all x.

We recall that a family of functions $\{F_n\}$ are said to be equicontinuous on $[a,b]$ if for every $\epsilon > 0$ there is a $\delta > 0$ (independent of n) such that whenever $|x-y| < \delta$

$$|F_n(x) - F_n(y)| < \epsilon \qquad \text{for all } n.$$

Corollary 7.9. Theorem 7.6 holds with condition (iii) in the controlled convergence replaced by : F_n are equicontinuous on $[a,b]$.

Proof. Suppose F_n are equicontinuous with $F_n(a) = 0$ for all n. We claim that $\{F_n(x)\}$ is uniformly bounded on $[a,b]$. Indeed, in view of equicontinuity, for every $x \in [a,b]$ and $\epsilon > 0$ there exists a $\delta(x) > 0$ such that

$$|F_n(x) - F_n(y)| \le \epsilon \quad \text{for every } n$$

whenever $|x - y| < \delta(x)$. We may put $\epsilon = 1$. Then it follows from the Heine-Borel covering theorem that there exists a finite number of points, say, $x_1 < x_2 < \ldots < x_N$, such that the union of $(x_i-\delta(x_i), x_i+\delta(x_i))$, $i = 1,2,\ldots,N$, covers $[a,b]$. For any $y \in [a,b]$ we have $y \in (x_i-\delta(x_i), x_i + \delta(x_i))$ for some i and, writing $x_o = a$,

$$\begin{aligned} |F_n(y)| &\le |F_n(y) - F_n(x_i)| + \sum_{k=1}^{i} |F_n(x_k) - F_n(x_{k-1})| \\ &\le 1 + 2i \\ &\le 2N + 1. \end{aligned}$$

Hence $\{F_n(x)\}$ is uniformly bounded. By Ascoli's theorem (which states that if $\{F_n(x)\}$ is uniformly bounded and equicontinuous then it has a uniformly convergent subsequence) the above sequence $\{F_n\}$ has a uniformly convergent subsequence. In view of the controlled convergence theorem, f is Henstock integrable and this subsequence converges to F, the primitive of f. Consequently, for every subsequence of $\{f_n\}$, there exists a sub-subsequence which converges uniformly to F on $[a,b]$. Therefore the sequence $\{F_n\}$ itself converges to F. That is, the consequence of Theorem 7.6 holds.

Corollary 7.10. If a function f is Henstock integrable on $[u,b]$ for each $u \in (a,b)$ and

$$\lim_{u\to a} \int_a^b f(x)dx = A \text{ exists,}$$

then f is Henstock integrable to A on $[a,b]$.

Proof. Let $f_n(x) = f(x)$ when $a + 1/n \le x \le b$ and 0 otherwise. Then the result follows from Corollary 7.7.

Corollary 7.11. Let X be a closed set in $[a,b]$ and $(a,b) - X$ the union of (c_k,d_k), $k = 1, 2, \ldots$. If f is Henstock integrable on X and on each $[c_k,d_k]$ with

$$\sum_{k=1}^{\infty} \omega(F;[c_k,d_k]) < +\infty$$

where F denotes the primitive of f over $[c_k,d_k]$, then f is Henstock integrable on [a,b] and

$$\int_a^b f = \int_X f + \sum_{k=1}^{\infty} \int_{c_k}^{d_k} f.$$

Proof. Define $f_n(x) = f(x)$ when $x \in X$ or $x \in [c_k,d_k]$ for $k = 1, 2, \ldots, n$ and 0 elsehwere. It is easy to verify that the conditions in the controlled convergence theorem are satisfied. Hence f is Henstock integrable on [a,b] and the equality holds.

We remark that Corollary 7.10 is known as the Cauchy extension and Corollary 7.11 the Harnack extension. In other words, the Henstock integral is closed under the Cauchy and Harnack extensions.

Corollary 7.12. If f is Henstock integrable on [a,b] then so is the truncated function f^N where $f^N(x) = f(x)$ when $|f(x)| \le N$, $f^N(x) = N$ when $f(x) > N$ and $f^N(x) = -N$ when $f(x) < -N$.

Proof. Let [a,b] be the union of X_1, X_2, ... such that the primitive F of f is $AC*(X_i)$ for each i. Define $f_n(x) = f(x)$ when $x \in X_1 \cup \ldots \cup X_n$ and 0 otherwise. Since F is $VB*(X_1 \cup \ldots \cup X_n)$ and, by Lemma 6.18, f_n is absolutely Henstock integrable on [a,b]. Therefore by Lemma 5.4 the truncated function f_n^N is also Henstock integrable. Apply the controlled controlled convergence theorem f^N as the limit function of f_n^N is Henstock integrable on [a,b].

We remark that Corollary 7.12 is a special case of Theorem 5.11 if we use the measurability of f.

8. THE PERRON INTEGRAL

We shall prove the equivalence of the Perron and Henstock integrals, and the generalized dominated convergence theorem.

Definition 8.1. A function H is said to be a major function of f on [a,b] if

$$-\infty \neq \underline{D}H(x) \geq f(x) \text{ for every } x$$

where $\underline{D}$ denotes the lower derivative. A function G is said to be a minor function of f on [a,b] if $-G$ is a major function of $-f$ on [a,b].

Definition 8.2. A function f is said to be Perron integrable on [a,b] if f has both major and minor functions and

$$-\infty < \inf\{H(b) - H(a)\} = \sup\{G(b) - G(a)\} < +\infty$$

where the infimum is over all major functions H of f on [a,b] and the supremum over all minor functions G of f on [a,b]. The common value is defined to be the Perron integral of f on [a,b].

It follows from the definition that a function f is Perron integrable on [a,b] if and only if for every $\epsilon > 0$ there exist a major function H and a minor function G such that $H(a) = G(a) = 0$ and

$$0 \leq H(b) - G(b) < \epsilon.$$

Hence the integral, if it exists, is uniquely determined.

Example 8.3. Let f be Newton integrable on [a,b]. Since its primitive F is both a major function and a minor function of f, then it is Perron integrable.

Example 8.4. Let $f(x) = 0$ almost everywhere in [a,b]. Then f is Perron integrable to 0 on [a,b]. A major function H is provided by the following lemma, and similarly for a minor function G.

Lemma 8.5. Let X be a set of measure zero in [a,b] and $\epsilon > 0$. Then there is a continuous and monotone increasing function H such that $H(a) = 0$, $H(b) < \epsilon$ and

$$H'(x) = +\infty \quad \text{for all } x \text{ in } X.$$

Proof. Given $\epsilon > 0$ and any positive integer n, we can construct an open set E_n which includes X and whose total length is less than $\epsilon 4^{-n}$. Define $h_n(x) = 2^n$ for $x \in E_n$ and 0 otherwise. Then write $H_n(x)$ to be the integral of h_n on [a,x] and further

$$H(x) = \sum_{n=1}^{\infty} H_n(x) \qquad \text{for } x \in [a,b],$$

which exists and satisfies all the conditions stated in the lemma.

Theorem 8.6. If H is a major function and G a minor function of f on [a,b], then the function H − G is monotone increasing on [a,b].

Proof. For any x, we have

$$\underline{D}(H(x) - G(x)) \geq \underline{D}\, H(x) + \underline{D}(-G(x)) \geq \underline{D}\, H(x) - \overline{D}\, G(x) \geq 0.$$

Hence H − G is monotone increasing.

Theorem 8.7. If f is Perron integrable on [a,b], then it is Henstock integrable there. Furthermore, they have the same integral value.

Proof. Since f is Perron integrable on [a,b], for every $\epsilon > 0$ there exist a major function H and a minor function G of f such that $H(a) = G(a) = 0$ and

$$0 \leq H(b) - G(b) < \epsilon.$$

For every $x \in [a,b]$, we have

$$f(x) - \epsilon < f(x) \leq \underline{D}H(x), \quad \overline{D}G(x) \leq f(x) < f(x) + \epsilon.$$

Then there exists a $\delta(x) > 0$ such that whenever $\xi \in [u,v] \subset (\xi-\delta(\xi), \xi+\delta(\xi))$ we have

$$f(\xi)(v - u) - \epsilon(v - u) \leq H(u,v),$$

$$G(u,v) \leq f(\xi)(v - u) + \xi(v - u),$$

where H(u,v) and G(u,v) denote H(v) − H(u) and G(v) − G(u) respectively.

Let F be the primitive of f and $D = \{[u,v];\xi\}$ any δ-fine division of [a,b]. Since both H − F and F − G are monotone increasing, therefore

$$G(u,v) \leq F(u,v) \leq H(u,v).$$

Then it follows that

$$\begin{aligned} G(u,v) - H(u,v) - \epsilon(v - u) &\leq F(u,v) - f(\xi)(v - u) \\ &\leq H(u,v) - G(u,v) + \epsilon(v - u). \end{aligned}$$

Summing over D, we obtain

$$\begin{aligned} |F(a,b) - \Sigma f(\xi)(v-u)| &\leq H(a,b) - G(a,b) + \epsilon(b-a) \\ &< \epsilon + \epsilon(b - a). \end{aligned}$$

Hence f is Henstock integrable on [a,b]. Obviously, they have the same integral value.

Theorem 8.8. If f is Henstock integrable on [a,b], then it is Perron integrable there.

Proof. Since f is Henstock integrable on [a,b], for every $\epsilon > 0$ there exists a $\delta(\xi) > 0$ such that for every δ-fine division D = $\{[u,v];\xi\}$ we have

$$\sum|f(\xi)(v - u) - F(u,v)| < \epsilon$$

where F(u,v) denotes the integral of f on [u,v]. Next define

$$\chi(x) = \sup \sum|f(\xi)(v - u) - F(u,v)|$$

where the supremum is over all δ-fine divisions D = $\{[u,v];\xi\}$ of [a,x]. Note that $\chi(a) = 0$ and $\chi(b) \le \epsilon$.

Write $\chi(x,y) = \chi(y) - \chi(x)$. For $\xi \in [u,v] \subset (\xi-\delta(\xi),\xi+\delta(\xi))$ we have

$$-\chi(u,v) \le f(\xi)(v-u) - F(u,v) \le \chi(u,v).$$

Dividing throughout by v − u, and taking limits, we obtain

$$\overline{D}(F(\xi) - \chi(\xi)) \le f(\xi) \le \underline{D}(F(\xi) + \chi(\xi))$$

for every $\xi \in [a,b]$ where $F(\xi)$ denotes $F(a,\xi)$. Hence we have shown that $F - \chi$ is a minor function and $F + \chi$ a major function of f in [a,b]. Furthermore,

$$0 \le (F(b) + \chi(b)) - (F(b) - \chi(b)) \le 2\epsilon.$$

The proof is complete.

We remark that we may define the Perron integral using only continuous major and minor functions. Indeed, we have the following

Theorem 8.9. A function f is Perron integrable on [a,b] if and only if for every $\epsilon > 0$ there exist a major function H and a minor function G, both continuous, such that H(a) = G(a) = 0 and

$$0 \le H(b) - G(b) < \epsilon.$$

Furthermore, the integral of f on [a,b] is the infimum of H(b) for all continuous major functions H with H(a) = 0.

We shall provide a proof of Theorem 8.9 at the end of Section 9.

To prove the generalized dominated convergence theorem, we need the following theorems.

Theorem 8.10. If the following condition is satisfied nearly everywhere in [a,b] (i.e., everywhere except perhaps for a countable number of points in [a,b]):

$$\overline{D}\, F(x) < +\infty \quad \text{or} \quad \underline{D}\, F(x) > -\infty,$$

then F is VBG* on [a,b].

Proof. If F is $VB^*(X)$ then it is $VB^*(X \cup \{c\})$ for any additional point c. So we may assume $\overline{D} F(x) < +\infty$ for all x and the other case is similar. Let X_{ni} denote the set of points $x \in [a + (i-1)/n, a + i/n]$ and $x \in [a,b]$ such that

$$(F(t) - F(x))/(t-x) \le n \quad \text{whenever } 0 < |t - x| \le 1/n.$$

Obviously, $[a,b]$ is the union of X_{ni} for $n, i = 1, 2, \ldots$.

Let a_{ni} and b_{ni} be respectively the infimum and the supremum of points in X_{ni}. For any two points $x, t \in X_{ni}$ with $x \le t$, we have

$$F(b_{ni}) - nb_{ni} \le F(t) - nt \le F(x) - nx \le F(a_{ni}) - na_{ni}.$$

Writing $F_n(x) = F(x) - nx$, we note that $F_n(x) - F_n(t) \ge 0$ when $x,t \in X_{ni}$ and $x \le t$. Furthermore,

$$\omega(F;[x,t]) \le \omega(F_n;[x,t]) + n(t - x) = F_n(x) - F_n(t) + n(t - x).$$

Consequently, for any partial division $D = \{[u,v]\}$ of $[a,b]$ with $u, v \in X_{ni}$ we have

$$\sum\omega(F;[u,v]) \le F_n(a_{ni}) - F_n(b_{ni}) + n(b_{ni} - a_{ni}).$$

That is, F is $VB^*(X_{ni})$. Therefore F is VBG^* on $[a,b]$.

We recall that a set X is said to be dense in Y if the closure $\overline{X} \supset Y$. A set of the form $Z \cap (x,y)$ is called a portion of Z.

Theorem 8.11. If X is closed and $X = \cup X_n$ then there is at least one X_n which is dense in a portion of X.

This is Baire's category theorem. We omit the proof. For example, if $[a,b]$ is the union of closed sets $X_1, X_2, \ldots$, then there is one X_n which contains some open interval. Again, if $X = \cup X_n$ and each X_n is closed, then there is one X_n and there is an interval (x,y) such that $X \cap (x,y) = X_n \cap (x,y)$. In particular, if F is VBG^* on $[a,b]$, then F is of bounded variation on some subinterval of $[a,b]$. Also, if F is VBG^* on a closed set X then there is a portion X_o of X such that F is $VB^*(X_o)$.

Now we are ready to prove the generalized dominated convergence theorem.

Theorem 8.12. If the following conditions are satisfied:

(i) $f_n(x) \to f(x)$ almost everywhere in $[a,b]$ as $n \to \infty$ where each

f_n is Henstock integrable on $[a,b]$;

(ii) f_n, $n = 1, 2, \ldots$, have at least one commmon major function H and at least one common minor function G in $[a,b]$;

(iii) the primitives F_n of f_n converge uniformly to a limit function F on $[a,b]$,

then f is Henstock integrable to $F(b) - F(a)$ on $[a,b]$.

Proof. We say that a point x is regular if the consequence of the theorem holds for some open subinterval containing x. Then the set Q of all points x not regular is closed. In view of Theorem 8.10 and 8.11, the set of regular points is nonempty.

Let (a_i,b_i), $i = 1, 2, \ldots$, be the subintervals of $[a,b]$ which are contiguous to Q. Then f is Henstock integrable to $F(v) - F(u)$ on every subinterval $[u,v]$ of (a_i,b_i). Since F is continuous in view of (iii), by Corollary 7.10 the function f is Henstock integrable on $[a_i,b_i]$ for every i.

Again, in view of Theorem 8.11, there is a portion Q_o of Q, i.e., Q_o is the intersection of Q and an open interval I_o for some I_o, such that H and G are $VB^*(Q_o)$. Take J_o to be the smallest interval that contains Q_o. Then it follows from Theorems 5.8 and 5.11 that f is Henstock integrable on the closure of Q_o. Also in view of the fact that

$$\omega(F_n;I) \le \omega(H;I) + \omega(G;I)$$

for any interval $I \subset J_o - Q_o$, the series of the oscillations of F_n over all intervals in J_o contiguous to Q_o converges uniformly in n. By Corollary 7.11, f is Henstock integrable on J_o which is a contradiciton. Hence the set Q is empty and the proof is complete.

We remark that the above proof by means of the category argument and by using the Cauchy extension (Corollary 7.10) and the Harnack extension (Corollary 7.11) is standard in the classical theory of the Denjoy-Perron integral. The technique also provides a second proof of the controlled convergence theorem if we prove the Cauchy and Harnack extensions independently.

Corollary 8.13. If a measurable function f has at least one continuous major function and at least one continuous minor function in

$[a,b]$, then f is Henstock integrable on $[a,b]$.

The proof is exactly the same as that of Theorem 8.12, in which the continuity of F at a_i and b_i is obtained by that of H and G and the following inequality

$$G(x,y) \le F(x,y) \le H(x,y) \qquad \text{for } [x,y] \subset (a_i,b_i).$$

The measurability of f is required when we apply Theorem 5.11. This is known as Marcinkiewicz' theorem. It is the nonabsolute version of the result that if a measurable function f is dominated by an absolutely Henstock integrable function g then f is also absolutely Henstock integrable. When the major and minor functions in Theorem 8.12 are continuous, then both $\sup\{f_n; n \ge 1\}$ and $\inf\{f_n; n \ge 1\}$ are Henstock integrable and the theorem reduces to the ordinary dominated convergence theorem (Theorem 4.3).

Again, let G and H be continuous on $[a,b]$. We see that if

$$\overline{D}\, G(x) \ge \underline{D}\, H(x) \qquad \text{for all } x,$$

then

$$G(a,b) \le \int_a^b \overline{D}\, G(x)dx \le \int_a^b \underline{D}\, H(x)dx \le H(a,b).$$

In other words, we may decompose G as follows : $G = G_a + G_s$ where G_a is continuous and ACG* and G_S is the singular part, and similarly for H.

Corollary 8.14. If, in addition, the major and minor functions in Corollary 8.13 are of bounded variation, then f is absolutely Henstock integrable on $[a,b]$.

This follows from Theorem 5.8. Alternatively, let $H^*(x)$ be the total variation of H on $[a,x]$ and $G^*(x)$ that of G on $[a,x]$. Then $H^* + G^*$ and $-H^* - G^*$ are respectively the continuous major and minor functions of $|f|$. Therefore f is absolutely Henstock integrable on $[a,b]$.

If f is Lebesgue integrable (see remark after Therorem 5.7), then f has a major function and a minor function, both of bounded variation. This provides another proof that every Lebesgue integrable function is Perron integrable.

Corollary 8.15. Let f_n be Henstock integrable on $[a,b]$ with the primitive F_n such that $f_n(x) \to f(x)$ almost everywhere in $[a,b]$ as

$n \to \infty$. If for every $\xi \in [a,b]$ and $\epsilon > 0$ there exists an integer N and a $\delta(\xi) > 0$ such that

$$|F_n(u,v) - F_m(u,v)| \leq \epsilon|v - u|$$

whenever $n, m \geq N$ and $\xi \in [u,v] \subset (\xi - \delta(\xi), \xi + \delta(\xi))$, then f is Henstock integrable on [a,b] and we have

$$\int_a^b f_n \to \int_a^b f \qquad \text{as} \quad n \to \infty$$

Proof. It follows easily from the condition that F converges uniformly to some F on [a,b]. Then the common major and minor functions are provided by $F(x) + \epsilon x$ and $F(x) - \epsilon x$. Hence the result.

We remark that the condition in the above corollary may hold only for $\xi \in [a,b] - D$ where D is countable and for each $\xi \in D$ the sequence F_n converges uniformly in an open neighbourbood of ξ.

Corollary 8.16. Let f_n be Henstock integrable on [a,b] with the primitive F_n such that $f_n(x) \to f(x)$ almost everywhere in [a,b] as $n \to \infty$. If F_n are uniformly differentiable in [a,b], i.e. for every $\xi \in [a,b]$ and $\epsilon > 0$ there exists a $\delta(\xi) > 0$ such that

$$|F_n(u,v) - f_n(\xi)(v-u)| \leq \epsilon|v - u|$$

whenever $\xi \in [u,v] \subset (\xi - \delta(\xi), \xi + \delta(\xi))$ and for all n, then the consequence of Corollary 8.15 holds.

The proof follows from the fact that the condition here is equivalent to that of Corollary 8.15.

As in the case of the Denjoy integral, we may again develop the Perron integral including the generalized dominated convergence theorem without reference to the Henstock integral. Of course, we will have to prove the Cauchy and Harnack extensions independently which is indeed possible.

We may define major and minor functions in a different form. In what follows, we consider a more general case, namely, the Stieltjes type.

Definition 8.17. A function H is called a Ward major function of f with respect to g in [a,b] if $H(a) = 0$ and for $\xi \in [a,b]$ there is a $\delta(\xi) > 0$ such that

$$f(\xi)\{g(v)-g(u)\} \le H(v) - H(u)$$

whenever $\xi - \delta(\xi) < u \le \xi \le v < \xi + \delta(\xi)$. A function G is called a Ward minor function of f with respect to g in [a,b] if $-G$ is a Ward major function of $-f$ in [a,b].

When $g(x) = x$, a Ward major function is a major function in the sense of Definition 8.1. Conversely, following the proof of Theorem 8.7 we see that $H(x) + \epsilon x$ is a Ward major function of f for $\epsilon > 0$.

Definition 8.18. A function f is said to be Ward integrable with respect to g on [a,b] if f has both Ward major and minor functions with respect to g in [a,b] and

$$-\infty < \inf H(b) = \sup G(b) < +\infty$$

where the infimum is over all Ward major functions H and the supremum over all Ward minor functions G. The common value A is defined to be the Ward integral of f with respect to g on [a,b], and we write

$$\int_a^b f(x)dg(x) = A.$$

The Ward integral is also known as the Perron-Stieltjes integral. When $\delta(\xi)$ is constant, it is called the Stieltjes integral. It is easy to see that when $g(x) = x$ the two versions of the Perron integrals coincide. Also, we may define correspondingly the Henstock integral of f with respect to g on [a,b] and show that it is equivalent to the Ward integral.

Example 8.19. If f is continuous and g of bounded variation on [a,b] then f is Ward integrable with respect to g on [a,b]. Following the proof of Example 2.6 with (b–a) replaced by the total variation $V(g;[a,b])$ we obtain that f is Henstock integrable with respect to g on [a,b]. In view of the previous remark, f is also Ward integrable with respect to g. Since f is uniformly continuous on [a,b] we may choose constant $\delta(\xi)$ and therefore the integral of f with respect to g is indeed the Stieltjes integral.

9. THE VARIATIONAL INTEGRAL

We shall prove in this section the generalized mean convergence theorem. Then use it to give a third proof of the controlled convergence theorem.

Definition 9.1. An interval function χ is said to be non-negative if $\chi(x,y) \geq 0$, and superadditive if

$$\chi(x,y) + \chi(y,z) \leq \chi(x,z) \qquad \text{when } x < y < z.$$

A real-valued function f is said to be variationally integrable on [a,b] with the primitive F if for every $\epsilon > 0$ there are a $\delta(\xi) > 0$ and a non-negative superadditive interval function χ such that

$$\chi(a,b) < \epsilon$$

and that whenever $\xi \in [u,v] \subset (\xi - \delta(\xi), \xi + \delta(\xi))$ we have

$$|F(u,v) - f(\xi)(v-u)| \leq \chi(u,v).$$

Here $F(u,v) = F(v) - F(u)$. As an example, if χ is a monotone increasing point function with $0 = \chi(a) \leq \chi(b) < \epsilon$, then $\chi(x,y) = \chi(y) - \chi(x)$ satisfies the condition in the definition.

Theorem 9.2. If f is variationally integrable on [a,b], then it is Henstock integrable there, and conversely.

Proof. Suppose f is variationally integrable on [a,b] and the condition in the definition holds. Then for any δ-fine division $D = \{[u,v];\xi\}$ we have

$$\begin{aligned} |F(a,b) - \sum f(\xi)(v-u)| &\leq \sum |F(u,v) - f(\xi)(v-u)| \\ &\leq \chi(a,b) < \epsilon. \end{aligned}$$

That is, f is Henstock integrable on [a,b].

Conversely, suppose f is Henstock integrable on [a,b]. Using the same notation, we put

$$\chi(x,y) = \sup \sum |f(\xi)(v-u) - F(u,v)|$$

where the supremum is over all δ-fine division $D = \{[u,v];\xi\}$ of [x,y]. Then χ satisfies the required condition and f is variationally integrable on [a,b].

Sometimes the results may be proved more easily using the variational integral. Note that the idea of the variational integral has already been used in the proof of Lemma 4.2 and that of Theroem 8.8.

Theorem 9.3. Let f_n, $n = 1, 2, \ldots$, be Henstock integrable on

$[a,b]$ with the primitives F_n, $n = 1, 2, \ldots$, $f_n(x) \to f(x)$ almost everywhere in $[a,b]$ as $n \to \infty$, and $F_n(x)$ converges pointwise to a limit function $F(x)$. Then in order that f is Henstock integrable on $[a,b]$ with the primitive F, it is necessary and sufficient that for every $\epsilon > 0$ there exists $M(\xi)$ taking integer values such that for infinitely many $m(\xi) \geq M(\xi)$ there are a $\delta(\xi) > 0$ and a non-negative superadditive interval function χ such that $\chi(a,b) < \epsilon$ and that whenever $\xi \in [u,v] \subset (\xi-\delta(\xi), \xi + \delta(\xi))$ we have

$$|F_{m(\xi)}(u,v) - F(u,v)| \leq \chi(u,v).$$

Proof. For simplicity, we may assume that $f_n(x) \to f(x)$ everywhere as $n \to \infty$. Suppose f is Henstock integrable on $[a,b]$ with the primitive F. Given $\epsilon > 0$ and $\xi \in [a,b]$, there is an integer $M(\xi)$ such that whenever $m(\xi) \geq M(\xi)$ we have

$$|f_{m(\xi)}(\xi) - f(\xi)| < \epsilon.$$

Since each f_n is also variationally integrable on $[a,b]$, there are a $\delta_n(\xi) > 0$ and a non-negative superadditive interval function χ_n such that $\chi_n(a,b) < \epsilon\, 2^{-n}$ and that for any δ_n-fine division $D = \{[u,v];\xi\}$ we have

$$|F_n(u,v) - f_n(\xi)(v-u)| \leq \chi_n(u,v).$$

Also, there are a $\delta_o(\xi) > 0$ and an interval function χ_o with $\chi_o(a,b) < \epsilon$ such that for any δ_o-fine division $D = \{[u,v];\xi\}$ we have

$$|F(u,v) - f(\xi)(v-u)| \leq \chi_o(u,v).$$

Now for every $m(\xi) \geq M(\xi)$, put $\delta(\xi) = \min\{\delta_{m(\xi)}(\xi), \delta_{o(\xi)}\}$ and

$$\chi(x,y) = \chi_o(x,y) + \sum_{n=1}^{\infty} \chi_n(x,y) + \epsilon(y-x).$$

Then it follows that whenever $\xi \in [u,v] \subset (\xi-\delta(\xi), \xi+\delta(\xi))$ we have

$$|F_{m(\xi)}(u,v) - F(u,v)| \leq |F_{m(\xi)}(u,v) - f_{m(\xi)}(\xi)(v-u)|$$
$$+ |f_{m(\xi)} - f(\xi)|(v-u) + |f(\xi)(v-u) - F(u,v)|$$
$$\leq \chi(u,v).$$

Hence we have proved the result for every $m(\xi) \geq M(\xi)$.

Conversely, suppose the condition is satisfied. Using the same notation, we choose $m(\xi) \geq M(\xi)$ such that

$$|f_{m(\xi)}(\xi) - f(\xi)| < \epsilon.$$

Then modify $\delta(\xi)$ so that $\delta(\xi) \le \delta_{m(\xi)}(\xi)$ for $\xi \in [a,b]$. Thus for any δ-fine division $D = \{[u,v];\xi\}$ we have

$$\begin{aligned}|F(u,v) - f(\xi)(v-u)| \le\ & |F(u,v) - F_{m(\xi)}(u,v)| \\ & + |F_{m(\xi)}(u,v) - f_{m(\xi)}(\xi)(v-u)| \\ & + |f_{m(\xi)}(\xi) - f(\xi)|(v-u) \\ \le\ & \chi(u,v) + \sum_{n=1}^{\infty} \chi_n(u,v) + \epsilon(v-u).\end{aligned}$$

Therefore, by definition, f is variationally integrable on [a,b] with the required interval function provided by the right side of the above inequality. Hence f is Henstock integrable on [a,b].

We remark that it is not enough for the condition in Theorem 9.3 to hold for a single $M(\xi)$ only. It must hold for infinitely many $m(\xi) \ge M(\xi)$ so that when proving the sufficiency we may choose a suitable $f_{m(\xi)}(\xi)$ which is sufficiently close to $f(\xi)$. Now the monotone convergence theorem follows easily from above by putting $\chi(u,v) = F(u,v) - F_p(u,v)$ for a sufficiently large p.

Corollary 9.4. Theorem 9.3 holds true with the necessary and sufficient condition replaced by : for every $\epsilon > 0$ there exists $M(\xi)$ taking integer values such that for infinitely many $m(\xi) \ge M(\xi)$ there is a $\delta(\xi) > 0$ such that for any δ-fine division $D = \{[u,v];\xi\}$ we have

$$|\sum F_{m(\xi)}(u,v) - F(a,b)| < \epsilon.$$

This is a re-statement of Theorem 9.3.

We shall now generalize the mean convergence theorem (Corollary 4.5). To motivate, let us rephrase the mean convergence of f_n in terms of their primitives F_n. More precisely, let f_n be Henstock integrable on [a,b] with the primitive F_n. If f_n is mean convergent then for every $\epsilon > 0$ there is a positive integer N such that whenever $n, m \ge N$ we have

$$\int_a^b |f_n - f_m| < \epsilon.$$

In view of Lemma 5.3, we may replace the above integral by the total variation of $F_n - F_m$ on $[a,b]$ or by the following sum

$$\sum\omega(F_n - F_m; [u,v]) < \epsilon$$

for any division $D = \{[u,v]\}$ of $[a,b]$ where Σ is over D and ω denotes the oscillation. Therefore it is natural to introduce a countable extension of the idea of the mean convergence as follows.

Definition 9.5. A sequence of functions $\{F_n\}$ is said to be oscillation convergent to F on $[a,b]$ if $[a,b]$ is the union of a sequence of closed sets X_i, $i = 1, 2, \ldots$, and for every i and $\epsilon > 0$ there is an integer N such that for any partial division $D = \{[u,v]\}$ of $[a,b]$ with all $u, v \in X$ we have

$$\sum \omega(F_n - F; [u,v]) < \epsilon \qquad \text{whenever } n \geq N.$$

LEMMA 9.6. If $\{F_n\}$ is oscillation convergent to F on $[a,b]$, then

(i) Definition 9.5 holds with $u, v \in X_i$ replaced by u or $v \in X_i$ for each i;

(ii) $[a,b]$ is the union of a sequence of closed sets X_i, $i = 1, 2, \ldots$, and for every i and $\epsilon > 0$ there are an integer N and a non-negative superadditive interval function χ such that $\chi(a,b) < \epsilon$ and whenever u or $v \in X_i$ we have

$$|F_n(u,v) - F(u,v)| \leq \chi(u,v) \qquad \text{for infinitely many } n \geq N.$$

Proof. To prove (i), we follow the same argument as in the proof of Lemma 6.15. We assume $a,b \in X_i$ for all i. Adopting the notation of Definition 9.5, take any partial division $D = \{[u,v]\}$ of $[a,b]$ with at least one of u and v belonging to X_i for each $[u,v]$ in D and for a fixed i. Sort out $[u,v]$ of D into two classes, one with $u \in X_i$ and another $v \in X_i$. If both $u, v \in X_i$ put $[u,v]$ in any one class. Then the sum of the oscillations over each class is less than ϵ. Hence

$$\sum \omega(F_n - F; [u,v]) < 2\epsilon \qquad \text{whenever } n \geq N.$$

Hence we have proved (i).

Next, fix X_i. In view of (i), for every $\epsilon > 0$ and every j there is an integer n(j) depending on i such that for any partial division $D = \{[u,v]\}$ of $[a,b]$ with u or $v \in X_i$ we have

$$\sum \omega(F_n - F;[u,v]) < \epsilon\, 2^{-j} \quad \text{whenever } n \ge n(j).$$

We may assume $n(j+1) \ge n(j)$ for all j. Let $N = n(1)$ and there are infinitely many $n \ge N$, namely, $n(j)$ for $j = 1, 2, \ldots$. We put

$$\chi_j(x,y) = \sup \sum \omega(F_{n(j)} - F;[u,v])$$

where the supremum is taken over all partial divisions $D = \{[u,v]\}$ of $[x,y]$ with u or $v \in X_i$ and Σ sums over D. Further, put

$$\chi(x,y) = \sum_{j=1}^{\infty} \chi_j(x,y).$$

Obviously, χ is non-negative and superadditive with $\chi(a,b) \le \epsilon$ and satisfying the required condition. Hence we have proved (ii).

Now we state and prove the generalized mean convergence theorem.

Theorem 9.7. If the following conditions are satisfied:

(i) $f_n(x) \to f(x)$ almost everywhere in $[a,b]$ as $n \to \infty$ where each f_n is Henstock integrable on $[a,b]$;

(ii) the primitives F_n of f_n are oscillation convergent to F on $[a,b]$;

(iii) the primitives F_n converge uniformly to F on $[a,b]$;

then f is Henstock integrable to $F(a,b)$ on $[a,b]$.

Proof. It suffices to show that the condition in Theorem 9.3 holds. Given $\epsilon > 0$, in view of Lemma 9.6(ii), for every i and j there exist $F_{n(i,j)}$ and a non-negative superadditive interval funciton χ_{ij} with $\chi_{ij}(a,b) < \epsilon 2^{-i-j}$ such that for any $[u,v]$ with u or $v \in X_i$ we have

$$|F_{n(i,j)}(u,v) - F(u,v)| \le \chi_{ij}(u,v).$$

We may assume that for each i, $\{F_{n(i,j)}\}$ is a subsequence of $\{F_{n(i-1,j)}\}$. Now consider $F_{n(j)} = F_{n(j,j)}$ in place of F_n and write $Y_1 = X_1$ and $Y_i = X_i - (X_1 \cup X_2 \cup \ldots \cup X_{i-1})$ for $i = 2, 3, \ldots$. Put $M(\xi) = n(i)$ when $\xi \in Y_i$. Note that there are infinitely many $m(\xi) \ge M(\xi)$, namely, all $n(j) \ge n(i)$.

For $n(j) \ge n(i)$, take arbitrary $\delta(\xi) > 0$ and define

$$\chi(u,v) = \sum_{i,j} \chi_{ij}(u,v).$$

Obviously, χ is non-negative, superadditive and $\chi(a,b) \le \epsilon$. Then whenever $\xi \in [u,v] \subset (\xi-\delta(\xi),\xi+\delta(\xi))$ and $\xi \in Y_i$ for some i we have

$$|F_{n(j)}(u,v) - F(u,v)| \le \chi_{ij}(u,v) \le \chi(u,v).$$

Apply Theorem 9.3 and we obtain that f is Henstock integrable to F(a,b) on [a,b].

Note that in the final step of the proof above, we take arbitrary $\delta(\xi) > 0$. Therefore Thereom 9.7 would still remain valid if we relax the oscillation convergence to that Definition 9.5 holds with

$$\sum\omega(F_n-F;[u,v]) < \epsilon \quad \text{whenever } n \ge N$$

for any δ_i-fine partial division $D = \{[u,v];\xi\}$ of [a,b] with $\xi \in X_i$ and for some $\delta_i(\xi) > 0$. Correspondingly, we can prove Lemma 9.6 with the added condition.

Theorem 9.8. If f_n is control-convergent to f on [a,b] then the primitives F_n of f_n are oscillation convergent to some function F on [a,b].

Proof. Let $[a,b] = \cup_i X_i$ such that the primitives F_n of f_n are $AC^*(X_i)$ uniformly in n for each i. Fix i and write $X = X_i$. We may assume X closed and write

$$(a,b) - X = \bigcup_{k=1}^{\infty} (c_k,d_k).$$

It follows from Lemma 6.18 that each $f_{n,X}$ is absolutely Henstock integrable on [a,b] with the primitive $F_{n,X}$ where $f_{n,X}(x) = f_n(x)$ when $x \in X$ and 0 otherwise. Also f_X is absolutely Henstock integrable on [a,b] with the primitive F_X by Theorem 7.1. We observe that

$$|F_{n,X}(u,v) - F_X(u,v)| \le \int_u^v |f_{n,X} - f_X|,$$

$$|F_n(u,v) - F_{n,X}(u,v)| \le \sum_{k=1}^{\infty} |F_n([u,v] \cap [c_k,d_k])|.$$

Since the above series converges uniformly in n, the last inequality still holds with F_n and $F_{n,X}$ replaced respectively by F and F_X.

Given $\epsilon > 0$, there is an integer p such that

$$\sum_{k=p+1}^{\infty} \omega(F_n;[c_k,d_k]) < \epsilon \qquad \text{for all } n.$$

Since, in addition, F_n converges uniformly to F on [a,b], there is an integer N such that for all $n \geq N$

$$\int_a^b |f_{n,X} - f_X| < \epsilon \qquad \text{and} \qquad \sum_{k=1}^{p} \omega(F_n - F; [c_k, d_k]) < \epsilon.$$

Take any partial division $D = \{[u,v]\}$ of [a,b] with $u, v \in X$. Let E_1 be the union of $[c_k, d_k]$ for $k = 1, 2, \ldots, p$. Then

$$\sum \omega(F_n - F; [u,v] \cap E_1) \leq \sum_{k=1}^{p} \omega(F_n - F; [c_k, d_k])$$

where $\sum$ sums over D. Let E_2 be the closure of $[a.b] - E_1$. Then

$$\begin{aligned}\sum\omega(F_n - F; [u,v] \cap E_2) &\leq \sum\omega(F_n - F_{n,X}; [u,v] \cap E_2) \\ &\quad + \sum\omega(F_{n,X} - F_X; [u,v] \cap E_2) + \sum\omega(F_X - F; [u,v] \cap E_2) \\ &\leq \sum_{k=p+1}^{\infty} \omega(F_n; [c_k, d_k]) + \int_a^b |f_{n,X} - f_X| \\ &\quad + \sum_{k=p+1}^{\infty} \omega(F; [c_k, d_k]).\end{aligned}$$

Combining all the above inequalities, we obtain

$$\begin{aligned}\sum\omega(F_n - F; [u,v]) &\leq \sum\omega(F_n - F; [u,v] \cap E_1) + \sum\omega(F_n - F; [u,v] \cap E_2) \\ &< 4\epsilon.\end{aligned}$$

That is, F_n are oscillation convergent to F on [a,b]. Hence we have proved the theorem.

Proof of the controlled convergence theorem (Theorem 7.6). It follows from Theorem 9.8 that the conditions in Theorem 9.7 are satisfied. Hence f is Henstock integrable on [a,b] and the integrals of f_n on [a,b] converge to that of f as $n \to \infty$.

Note that the above proof depends heavily on a real-line property, namely, an open set on the real line is the union of a countable number of open intervals. A proof which is real-line independent and which follows roughly the same procedure (using Corollary 9.4 in place of Theorem 9.3) can be found in Section 21.

We end this section with the following proof.

Proof of Theorem 8.9. We prove only the necessity. We apply the category argument as in the proof of Theorem 8.12. We say that a point x is regular if the condition in Theorem 8.9 holds on some open subinterval containing x. That is, we can choose the major and minor functions to be continuous. Then the set Q of all points x not regular is closed. Again, the set of regular points is nonempty.

We observe from the proof of Theorem 8.7 that if a function f is Perron integrable using only continuous major and minor functions H and G respectively then it is variationally integrable with $\chi(x) = H(x) - G(x) + \epsilon x$ which is also continuous. In other words, throughout the proof of the controlled convergence theorem using the variational integral we may consider continuous χ functions. Hence we have proved the controlled convergence theorem for the Perron integral using only continuous major and minor functions. In particular, the Cauchy extension (Corollary 7.10) and the Harnack extension (Corollary 7.11) hold for the Perron integral using only continuous major and minor functions.

The rest of the proof follows in exactly the same way as that of Theorem 8.12. Hence Q is empty and the Perron integral using continuous major and minor functions is equivalent to the original definition (Definition 8.2).

10. A RIESZ-TYPE DEFINITION

Riesz defined the Lebesgue integral by means of a sequence of step functions satisfying certain condition, namely, the mean convergence. In what follows, we give a Riesz-type definition for the Henstock integral.

Definition 10.1. A function f is said to be RD integrable on [a,b] if there is a control-convergent sequence $\{\psi_n\}$ of step functions such that $\psi_n(x) \to f(x)$ almost everywhere in [a,b] as $n \to \infty$. We define

$$\int_a^b f(x)dx = \lim_{n\to\infty} \int_a^b \psi_n(x)dx.$$

We see from the controlled convergence theorem that the above integral exists and is uniquely determined. It is obvious that if f is RD integrable on [a,b] then it is also Henstock integrable there.

In fact, they are equivalent.

Theorem 10.2. If f is Henstock integrable on [a,b], then it is RD integrable there.

Proof. Suppose f is Henstock, or equivalently, Denjoy integrable on [a,b]. Then the primitive F of f is ACG* on [a,b], i.e., [a,b] is the union of closed sets X_i, i = 1, 2, ..., on each of which F is $AC^*(X_i)$. For convenience, assume $a,b \in X_i$ for all i. Let $F_n(x) = F(x)$ when $x \in X_1 \cup \ldots \cup X_n$ and linear elsewhere. Put $f_n(x) = F_n'(x)$ almost everywhere. Then either the sequence $\{f_n\}$ itself or a subsequence $\{f_{n(i)}\}$ of $\{f_n\}$ is control-convergent to f on [a,b]. For each n, by Theorem 5.9 there is step function ψ_n satisfying the following condition:

$$\int_a^b |f_n - \psi_n| < 2^{-n}.$$

It follows from the monotone convergence theorem (Theorem 4.1) that

$$\sum_{n=1}^{\infty} |f_n(x) - \psi_n(x)| \quad \text{converges almost everywhere}$$

and is Henstock integrable on [a,b]. In other words, $\psi_n(x) = \{\psi_n(x) - f_n(x)\} + f_n(x)$ converges to f(x) almost everywhere in [a,b] as $n \to \infty$. In view of the fact that

$$\left|\int_u^v \psi_n\right| \le \int_u^v |f_n - \psi_n| + \left|\int_u^v f_n\right|$$

either the sequence $\{\psi_n\}$ itself or a subsequence $\{\psi_{n(i)}\}$ of $\{\psi_n\}$ is control-convergent to f on [a,b]. Hence f is RD integrable on [a,b].

We may develop the theory of the integral using Definition 10.1. First, we have to prove the uniqueness. The proof is in fact a special case of the controlled convergence theorem. If done, then we may use it to prove the general controlled convergence theorem as follows.

Proof of the controlled convergence theorem. Given that the sequence $\{f_n\}$ is control-convergent to f on [a,b], therefore the primitives F_n of f_n are ACG* on [a,b] uniformly in n, i.e., [a,b] is the union of closed sets X_1, X_2, ... on each of which F_n is $AC^*(X_i)$ uniformly in n. Again, assume $a,b \in X_i$ for all i. Put $G_n(x) = F_n(x)$

when $x \in X_1 \cup \ldots \cup X_n$ and linear elsewhere, and $g_n(x) = G_n'(x)$ almost everywhere. Following the same idea as in the proof of Theorem 10.2, we can construct, for each n, a step function ψ_n such that

$$\int_a^b |g_n - \psi_n| < 2^{-n}.$$

Again we can show that either the sequence $\{\psi_n\}$ itself or a subsequence $\{\psi_{n(i)}\}$ of $\{\psi_n\}$ is control-convergent to f on [a,b]. Hence f is RD integrable on [a,b] and

$$\int_a^b f = \lim_{n(i)\to\infty} \int_a^b \psi_{n(i)}$$

$$= \lim_{n(i)\to\infty} \int_a^b g_{n(i)}$$

$$= \lim_{n(i)\to\infty} \int_a^b f_{n(i)}.$$

Since every subsequence of $\{f_n\}$ has a sub-subsequence which satisfies the required condition, so that sequence $\{f_n\}$ itself satisfies the condition. The proof is complete.

As an application of the RD integral, we prove the following

Theorem 10.3. If f is Henstock integrable to F(a,b) on [a,b], then for every $\epsilon > 0$ there is a measurable function $\delta(\xi) > 0$ such that for any δ-fine division $D = \{[u,v]; \xi\}$ of [a,b] we have

$$|\sum f(\xi)(v-u) - F(a,b)| < \epsilon.$$

Proof. We shall construct a positive function $\delta(\xi)$ which is measurable and satisfies the required conditon. It follows from Theroem 10.2 that there is a sequence of step functions $\{f_n\}$ which is control-convergent to f on [a,b]. It follows from Theorem 9.8 that if $\{f_n\}$ is control-convergent to f then the primitives F_n of f_n are oscillation convergent to a function F. Further, in the proof of Theorem 9.7, we have shown that the conditon in Theorem 9.3 holds and therefore that in Corollary 9.4. More precisely, for every $\epsilon > 0$ there exists $M(\xi)$ taking integer values such that for infinitely many $m(\xi) \geq M(\xi)$ and for any partial division $D = \{[u,v];\xi\}$ of [a,b] with

$\xi = u$ or v we have

$$|\sum F_{m(\xi)}(u,v) - F(a,b)| < \epsilon.$$

Note that no $\delta(\xi)$ is involved so far. Also, we may choose $M(\xi)$ and $m(\xi)$ as follows. We can find a sequence of positive integers $n(1)$, $n(2)$, ... and a sequence of closed sets $X_1, X_2, \ldots$ such that $M(\xi) = n(i)$ when $\xi \in Y_i$ where $Y_1 = X_1$ and $Y_i = X_i - (X_1 \cup \ldots \cup X_{i-1})$ for $i = 1,2,3, \ldots$, and that $m(\xi)$ takes integer values from $\{n(j); n(j) \geq n(i)\}$ when $\xi \in Y_i$.

Next, since each f_n is a step function, given $\epsilon > 0$ there is a constant $\delta_n > 0$ such that for any δ_n-fine division $D = \{[u,v];\xi\}$ we have

$$|\sum f_n(\xi)(v-u) - F_n(a,b)| < \epsilon\, 2^{-n}.$$

Relabelling, if necessary, we may assume $n(i) = i$. Let $f_n(x) \to f(x)$ everywhere as $n \to \infty$ except perhaps in a set S of measure zero. For $\xi \in [a,b] - S$, define $j(\xi)$ to be the minimum value of all j such that

$$|f_k(\xi) - f(\xi)| < \epsilon \qquad \text{whenever } k \geq j.$$

When $j(\xi) > 1$, we have

$$|f_k(\xi) - f(\xi)| < \epsilon \quad \text{whenever} \quad k \geq j(\xi),$$

$$|f_{j(\xi)-1}(\xi) - f(\xi)| \geq \epsilon.$$

This is the crucial step in the proof, the rest is the same as the proof of the controlled convergence theorem. Note that $j(\xi)$ may be smaller than $M(\xi)$. Now define $\delta(\xi) = \delta_{j(\xi)}$ when $j(\xi) \geq i$ and $\delta(\xi) = \delta_i$ when $j(\xi) < i$ for $\xi \in Y_i - S$ and $i = 1, 2, \ldots$. It remains to define $\delta(\xi)$ on S.

As in the proof of Theorem 6.12, let $\delta(\xi)$ be defined on S so that for any δ-fine partial division $D = \{[u,v];\xi\}$ of $[a,b]$ with $\xi \in S$ we have

$$|\sum f(\xi)(v - u)| < \epsilon, \qquad |\sum F(u,v)| < \epsilon,$$

where $\sum$ sums over D. Now $\delta(\xi)$ is defined on the whole interval $[a,b]$. Then for any δ-fine division $D = \{[u,v];\xi\}$ of $[a,b]$ we have

$$|\sum f(\xi)(v-u) - F(a,b)| \leq |\sum_1 \{f(\xi) - f_{j(\xi)}(\xi)\}(v-u)|$$

$$+ |\sum_1 \{f_{j(\xi)}(v-u) - F_{j(\xi)}(u,v)\}|$$

$$+ \left| \sum_1 \{F_{j(\xi)}(u,v) - F(u,v)\} \right| + \left|\sum_2 f(\xi)(v-u)\right| + \left|\sum_2 F(u,v)\right|$$
$$< \epsilon(b-a+4)$$

where $\sum$ sums over D and $\sum = \sum_1 + \sum_2$ in which $\sum_1$ sums over all $\xi \notin S$ and $\sum_2$ over $\xi \in S$. That is, f is Henstock integrable on [a,b] with the given $\delta(\xi)$. It remains to show that $\delta(\xi)$ so defined is measurable.

Since S is of measure zero, it suffices to show that $\delta(\xi)$ is measurable on [a,b] − S. Fix i and let E_p denote the set of all $\xi \in Y_i - S$ such that

$$|f_k(\xi) - f(\xi)| < \epsilon \quad \text{whenever } k \geq p; \quad |f_{p-1}(\xi) - f(\xi)| \geq \epsilon.$$

Obviously, E_p is a measurable set by Theorem 5.12. Note that $\delta(\xi)$ takes constant value δ_p on E_p. On the other hand, $\delta(\xi)$ takes constant value δ_i on the measurable set

$$(Y_i - S) - \bigcup_{p \geq i} E_p.$$

Therefore by Theorem 5.12 again $\delta(\xi)$ is a measurable function on $Y_i - S$ and consequently on [a,b].

Now we discuss the relationship among the various convergence theorems. A sequence of functions $\{f_n\}$ is said to be generalized mean convergent to f on [a,b] if the conditions in the generalizd mean convergence theorem (Theorem 9.7) are satisfied. Hence it follows from Theorem 9.8 that the controllled convergence implies the generalized mean convergence.

A sequence of functions $\{f_n\}$ is said to be generalized dominately convergent to f on [a,b] if the conditions in the generalised dominated convergence theorem (Theorem 8.12) are satisfied. Then we can prove the following

Theorem 10.4. If a sequence $\{f_n\}$ is generalized dominately convergent to f on [a,b], then it is control-convergent to f there.

Proof. It suffices to show that the primitives F_n of f_n are ACG* uniformly in n. The proof follows that of Theorem 6.19. First, we observe that by Theorem 8.10 both the common major function H and the common minor function G are VBG*. Since $G(u,v) \leq F_n(u,v) \leq H(u,v)$ for all [u,v] and all n, F_n are VBG* uniformly in n. In other words,

$[a,b] = \cup X_i$ such that F_n are $VB^*(X_i)$ uniformly in n for each i. In view of Lemma 6.4, it remains to show that F_n are $AC(X_i)$ uniformly in n for each i.

Fix i and write $X = X_i$. As in the proof of Theorem 6.19, let $f_{n,X}(x) = f_n(x)$ when $x \in X$ and zero otherwise and $F_{n,X}$ the primitive of $f_{n,X}$ on [a,b]. Then we have

$$F_n(x) = F_{n,X}(x) + \sum_{k=1}^{\infty} F_n([a,x] \cap [c_k,d_k]) \text{ for } x \in [a,b]$$

where (c_k,d_k) k = 1, 2, ..., are the intervals contiguous to X and the series converges uniformly in n. In the same way, we can show that F_n are AC(X) uniformly in n. The proof is complete.

Theorem 10.5. If $\{f_n\}$ is generalized mean convergent to f on [a,b] then there is a subsequence $\{f_{n(i)}\}$ of $\{f_n\}$ which is control-convergent to f on [a,b].

Proof. Suppose the primitives F_n of f_n are oscillation convergent on [a,b] with $X_1, X_2, \ldots$ given as in Definition 9.5. Consider X_1 and let $[a_k,b_k]$, k = 1,2,..., denote the subintervals of [a,b] contiguous to X_1. We may assume $a,b \in X_1$. Now define $G_n(x) = F_n(x)$ when $x \in X_1$ and piecewise linear on the complement of X_1 such that

$$\omega(G_n;[a_k,b_k]) = \omega(F_n;[a_k,b_k]) \quad \text{for all } k.$$

Let $g_n(x) = G_n'(x)$ almost everywhere. Then g_n is mean convergent on [a,b], i.e.

$$\int_a^b |g_n-g_m| \to 0 \quad \text{as } n,m \to \infty.$$

Hence as in the proof of Corollary 4.5 there is a subsequence $\{g_{n(i)}\}$ of $\{g_n\}$ which is dominated on the right by $g_{n(1)}+h$ and on the left by $g_{n(1)}-h$ where

$$h(x) = \sum_{i=1}^{\infty} |g_{n(i)}(x)-g_{n(i+1)}(x)|.$$

Furthermore, $g_{n(i)}(x)$ converges almost everywhere to a Henstock integrable function g(x) on [a,b]. Note that $g_{n(i)}(x) = f_{n(i)}(x)$ almost everywhere in X_1. Hence $f_{n(i)}(x)$ converges almost everywhere in X_1 to a function f(x) where f(x) = g(x) when $x \in X_1$.

Since $f_{n(1)}$ is Henstock integrable on [a,b] and its primitive $F_{n(1)}$ is ACG^*, then [a,b] is the union of closed sets Y_j, $j = 1,2,\ldots$, such that $F_{n(1)}$ is $AC^*(Y_j)$ for each j. In view of $g_{n(i)}$ being dominated by $g_{n(1)}+h$ and $g_{n(1)}-h$, we see that the functions $G_{n(i)}$ are $AC^*(Y_j)$ uniformly in n(i) and therefore $F_{n(i)}$ are $AC^*(X_1 \cap Y_j)$ uniformly in n(i) for each j. In other words, $F_{n(i)}$ are ACG^* on X_1 uniformly in n(i).

Repeat the above process for X_2 and the sequence $\{F_{n(i)}\}$ obtained above in place of X_1 and F_n, and so on. Consequently, by the diagonal process, we obtain a new subsequence $\{f_{n(i)}\}$ of $\{f_n\}$ such that $f_{n(i)}(x) \to f(x)$ almost everywhere in [a,b] as $n(i) \to \infty$ and that the primitives are ACG^* on [a,b] uniformly in n(i). Hence $f_{n(i)}$ is control-convergent to f on [a,b].

As we can see, the pointwise convergence of $f_n(x)$ is not required in the above proof. In fact, we have proved the following theorem.

Theorem 10.6. Let f_n be Henstock integrable on [a,b] and the primitives F_n of f_n are oscillation convergent and converge uniformly on [a,b]. Then there exists a subsequence $\{f_{n(i)}\}$ of $\{f_n\}$ such that $f_{n(i)}(x) \to f(x)$ almost everywhere in [a,b] as $n(i) \to \infty$. Furthermore,

$$\int_a^b f_n \to \int_a^b f \qquad \text{as } n\to\infty.$$

Example 10.7. Consider the sequence $\{f_n\}$ in Example 7.5, i.e., $f_n(x) = F'(x)$ for $1/n \le x \le 1$ and 0 elsewhere in [0,1], where $F(0) = 0$ and

$$F(x) = x^2 \sin x^{-2} \qquad \text{when } x \neq 0.$$

Define $H(x) = F(x)$ when $F'(x) \ge 0$ and 0 elsewhere in [0,1], and also $G(x) = F(x)$ when $F'(x) \le 0$ and 0 elsewhere in [0,1]. Then $\overline{D}G(x) \le f_n(x) \le \underline{D}H(x)$ nearly everywhere (i.e., everywhere except for a countable number of points) and for all n. In fact, the above inequality does not hold when $F'(x) = 0$. At those points which are countable we have $\overline{D}G(x) = +\infty$ and $\underline{D}H(x) = -\infty$. Hence we fail to find common major and minor functions in the sense of Definition 8.1 for the sequence $\{f_n\}$ or any of its subsequences though F_n are ACG^* uniformly in n.

We observe from the above example that the condition of generalized dominated convergence is still too restrictive, and should be relaxed further. However the technique of proof in Theorem 8.12 is an important tool. It is repeated when proving Corollary 8.13, Theorem 8.9 and later Theorem 17.8.

11. SARGENT SPACES

We shall introduce a special kind of normed linear spaces, called Sargent spaces. Then we prove the Riesz representation theorems for linear functionals (Section 12) and nonlinear functionals (Section 15) defined on a special Sargent space, namely, the space of all Henstock integrable functions on [a,b].

A normed linear space E is a linear space provided with a norm $\|x\|$ satisfying the following properties:

(i) $\|x\| \geq 0$ and $\|x\| = 0$ if and only if $x = 0$;

(ii) $\|x + y\| \leq \|x\| + \|y\|$;

(iii) $\|\alpha x\| = |\alpha|\ \|x\|$ for real α.

A normed linear space E is complete if $\|x_n - x_m\| \to 0$ as $n, m\to\infty$ implies that there is $x \in E$ such that $\|x_n - x\| \to 0$ as $n \to \infty$, i.e., every Cauchy sequence is convergent. A complete normed linear space is called a Banach space. For example, the space of all continuous functions on [a,b] provided with the uniform norm $\|f\| = \sup\{|f(x)|;\ x \in [a,b]\}$ is a Banach space.

Example 11.1. Let E be the space of all Henstock integrable functions on [a,b]. We define a norm in E as follows:

$$\|f\| = \sup\{\left|\int_a^x f(t)dt\right|;\ a \leq x \leq b\}.$$

As usual, we regard two functions f and g as identical if $f(x) = g(x)$ almost everywhere in [a,b]. Then E is a normed linear space and we call it the Denjoy space.

We shall show that E is not a Banach space. Let $[a,b] = [0,1]$ and K be the Cantor ternary set in [0,1]. Then the complement of K is the following set of open intervals:

$$(\tfrac{1}{3}, \tfrac{2}{3}), (\tfrac{1}{3^2}, \tfrac{2}{3^2}), (\tfrac{2}{3} + \tfrac{1}{3^2}, \tfrac{2}{3} + \tfrac{2}{3^2}), (\tfrac{1}{3^3}, \tfrac{2}{3^3}), (\tfrac{2}{3^2} + \tfrac{1}{3^3}, \tfrac{2}{3^2} + \tfrac{2}{3^3}),$$

$$(\tfrac{2}{3} + \tfrac{1}{3^3}, \tfrac{2}{3} + \tfrac{2}{3^3}), (\tfrac{2}{3} + \tfrac{2}{3^2} + \tfrac{1}{3^3}, \tfrac{2}{3} + \tfrac{2}{3^2} + \tfrac{2}{3^3}), \ldots$$

For convenience we shall label these intervals respectively $I(1/2)$, $I(1/4)$, $I(3/4)$, $I(1/8)$, $I(3/8)$, $I(5/8)$, $I(7/8)$, ...

Let $F_1(0) = 0$, $F_1(1) = 1$, and $F_1(x) = 1/2$ when $x \in I(1/2)$. On each of the two components of $[0,1] - I(1/2)$ let F_1 be a linear segment drawn in such a way that the function F_1 is continuous on $[0,1]$.

Let $F_2(x) = F_1(x)$ when $x = 0$, $x = 1$ or $x \in I(1/2)$. On each of the intervals $I(1/4)$, $I(3/4)$ let F_2 be constant and be equal to $1/4$ and $3/4$ respectively. On each of the four components of $[0,1] - \{I(1/2) \cup I(1/4) \cup I(3/4)\}$ let F_2 be a linear segment drawn in such a way that F_2 is continuous on $[0,1]$.

This procedure can be continued to give a sequence of absolutely continuous functions $\{F_n\}$ defined on $[0,1]$. The sequence $\{F_n\}$ so defined is uniformly convergent on $[0,1]$ and tends to a limit function F. Since K is of measure zero and yet $F(K) = \{f(x); x \in K\}$ is of measure 1, therefore F cannot be ACG^* on $[0,1]$.

Let $f_n(x) = F'_n(x)$ for almost all $x \in [0,1]$. Then f_n is a Cauchy sequence in E and has no limit in E. Therefore E is not complete.

We recall that a set X is said to be dense in Y if the closure $\overline{X} \supset Y$. A ball B in a normed linear space E is the set of all $x \in E$ such that $\|x - x_o\| \leq r$ for some $x_o \in E$ and some real number r. If a set X is not dense in any ball B in the space E, then we say : X is nowhere dense in E. In other words, for any ball B in E there is a smaller ball B_o contained in B such that B_o does not intersect X.

Definition 11.2. A sequence $\{X_n\}$ of sets is an α-sequence if $0 \in X_1$ and for every n, $x + y$ and $x - y \in X_{n+1}$ whenever $x, y \in X_n$. A normed linear space E is called an α-space if

$$E = \bigcup_{n=1}^{\infty} X_n$$

where $\{X_n\}$ is an α-sequence of closed sets each of which is nowhere dense in E. A normed linear space is called a Sargent space or a β-space if it is not an α-space.

We recall that a space is said to be of the first category if it is a countable union of nowhere dense sets. If it is not of the first category, then we say it is of the second category. Obviously, an α-space is of the first category, whereas a space of the second category is necessarily a Sargent space.

Example 11.3. The Denjoy space E as given in Example 11.1 is a Sargent space.

We sketch the proof as follows. Suppose E is not a Sargent space. Then we have

$$E = \bigcup_{n=1}^{\infty} Q_n$$

where $\{Q_n\}$ is an α-sequence of closed sets each of which is nowhere dense in E. Denote by E(c,d) the class of functions in E defined on $[c,d] \subset [a,b]$ and zero outside [c,d], and similarly for $Q_n(c,d)$. An element in E(c,d) will be denoted by $x_{c,d}$. If $Q_n(c,d)$ is nowhere dense in E(c,d) for every n, and $c < e < d$, then either $Q_n(c,e)$ is nowhere dense in E(c,e), $n = 1,2,\ldots$, or $Q_n(e,d)$ is nowhere dense in E(e,d), $n = 1,2,\ldots,$. It follows that there is a decreasing sequence of nested intervals $[a_s,b_s]$, $s = 1,2,\ldots$, such that their intersection consists of a single point and that for each s, the sequence $\{Q_n(a_s,b_s)\}$ is an α-sequence of closed sets each of which is nowhere dense in $E(a_s,b_s)$. Then there are two possiblities which may both occur:

(i) There exists a point c and an increasing sequence $\{c_s\}$ having c as its limit and such that $\{Q_n(c_s,c)\}$ is an α-sequence of closed sets each of which is nowhere dense in $E(c_s,c)$.

(ii) There exists a point d and a decreasing sequence $\{d_s\}$ having d as its limit and such that $\{Q_n(d,d_s)\}$ is an α-sequence of closed sets each of which is nowhere dense in $E(d,d_s)$.

For definiteness, we assume (i). We shall prove by induction that there exist a subsequence $\{h_n\}$ of $\{c_s\}$ and a sequence of closed balls $S_1,S_2,\ldots$ with centres $x_1,x_2,\ldots$ and radii $r_1,r_2,\ldots$ such that for every positive integer n

(*) $\quad S_n \subset S_{n-1}, \quad S_n \cap Q_n = 0, \quad r_n < 2^{-n}, \quad x_n - x_{n-1} \in E(h_n,h_{n+1})$

where it is understood that $S_0 = E$ and $x_0 = 0$.

Since $Q_1(c_1,c)$ is nowhere dense in $E(c_1,c)$, there is $y \in E(c_1,c)$ such that $y \notin Q_1(c_1,c)$. Since $Q_1(c_1,c)$ is closed and $\lim_{n\to\infty} y_{c_1,c_n} = y$, there is a number $m > 1$ such that $y_{c_1,c_m} \notin Q_1(c_1,c)$, for otherwise y would belong to $Q_1(c_1,c)$. It follows that there is $x_1 \in E(c_1,c_m)$ such that $x_1 \notin Q_1(c_1, c)$ and hence $x_1 \notin Q_1$. Therefore taking $h_1 = c_1$, $h_2 = c_m$ we can choose a closed ball S_1 with centre x_1 and radius r_1 such that the conditions in (*) are satisfied for $n = 1$.

Suppose that the statement is true for $n = 1, 2, \ldots, m$. Then $x_m \notin Q_m$ implies $x_m \in Q_p$ for some $p \geq m+1$. Since $Q_{p+1}(h_{m+1}, c)$ is nowhere dense in $E(h_{m+1}, c)$, there is $y \in E(h_{m+1}, c)$ such that $\|y\| \leq r_m$ and $y \notin Q_{p+1}(h_{m+1}, c)$. It follows by arguments similar to those used in determining h_2 and x_1 that there is $z \in E(h_{m+1}, h_{m+2})$ such that $\|z\| \leq r_m$ and $z \notin Q_{p+1}(h_{m+1}, c)$ and hence $z \notin Q_{p+1}$.

Set $x_{m+1} = x_m + z$. It is easy to see that x_{m+1} is an interior element of S_m and $x_{m+1} \notin Q_{m+1}$. Therefore we can choose a closed ball S_{m+1} with centre x_{m+1} and radius r_{m+1} such that the conditions in (*) hold for $n = m+1$. This proves the existence of (S_n) having the properties stated above.

We have proved (*). Since $x_n = \sum_{s=1}^{n} (x_s - x_{s-1})$, while $x_s - x_{s-1} \in E(h_s, h_{s+1})$ and

$$\sum_{s=1}^{\infty} \|x_s - x_{s-1}\| \leq \|x_1\| + \sum_{s=1}^{\infty} r_s < \infty,$$

it can be proved that the sequence (x_n) is convergent. Set $x = \lim_{n\to\infty} x_n$. Since S_m is closed for any fixed positive integer m and $x_n \in S_m$ whenever $n > m$, then $x \in S_m$. In view of $S_m \cap Q_m = 0$, we get $x \notin Q_m$ for all m. This contradicts the fact that $x \in \bigcup_{n=1}^{\infty} Q_n$. Hence E must be a Sargent space.

Example 11.4. Every Banach space is a Sargent space. Baire's category theorem states that every Banach space is of the second category. Hence, by the remark after Definition 11.2, it is a Sargent space. However not every Sargent space is of the second category. For example, the Denjoy space is indeed of the first category. We shall not prove this result. For a reference, see the bibliography.

Next, we consider continuous linear operators on Sargent spaces. Let T be an operator from a normed linear space into another. It is linear if

$$T(x + y) = T(x) + T(y).$$

It is continuous if $\|x_n - x\| \to 0$ as $n \to \infty$ implies that

$$\lim_{n\to\infty} \|T(x_n)-T(x)\| = 0.$$

It is well-known that a linear operator T is continuous if and only if it is bounded in the sense that

$$\|T(x)\| \le M\|x\| \quad \text{for all } x.$$

Then the norm $\|T\|$ of T is defined to be the infimum of all such M above. An operator taking real values is called a functional.

We shall show that the Banach-Steinhaus theorem still holds for Sargent spaces. First, we state a lemma, which is a direct consequence of the definition of Sargent spaces.

Lemma 11.5. A normed linear space E is a Sargent space if and only if for every representation of the form

$$E = \bigcup_{n=1}^{\infty} X_n$$

where $\{X_n\}$ is an α-sequence, there is an X_N for some N which is dense in a ball B in E.

The above may be used as an alternative definition of Sargent spaces. Lemma 11.5 plays the same role in the following theorem as Baire's category theorem in the proof of the Banach-Steinhaus theorem for Banach spaces.

Theorem 11.6. Let T_n be a sequence of continuous linear operators from a Sargent space E into a normed linear space F. If $\sup\{\|T_n(x)\|;\ n \ge 1\} < +\infty$ for every $x \in E$, then

$$\sup\{\|T_n\|;\ n \ge 1\} < +\infty.$$

Proof. Let Y_{nm} be the set of all $x \in E$ such that $\|T_n(x)\| \le 2^m$, and

$$X_m = \bigcap_{n=1}^{\infty} Y_{nm}.$$

Since T_n are continuous, the set Y_{nm} are closed and so are X_m. Furthermore, $\{X_m\}$ is an α-sequence.

It follows from Lemma 11.5 that there are an integer m and a ball with centre x_0 and radius r such that

$$\|T_n(x)\| \le 2^m$$

for all n and whenever $\|x-x_0\| \le r$. Then for any non-zero $x \in E$ we have

$$\left\|T_n\left(\frac{rx}{\|x\|}\right)\right\| \le \left\|T_n\left(\frac{rx}{\|x\|} + x_0\right)\right\| + \|T_n(x_0)\|$$

$$\le 2^m + 2^m.$$

Thus for all n and all $x \in E$

$$\|T_n(x)\| \le \frac{2^{m+1}}{r}\|x\|.$$

Hence $\{\|T_n\|\}$ is bounded.

To state the next theorem, we recall that a subset of E is said to be fundamental if the set of all linear combinations of elements in the subset is dense in E.

Theorem 11.7. Let T and T_n, $n = 1,2,\ldots$, be continuous linear operators on a Sargent space E into a normed linear space F. Then

$$\lim_{n\to\infty} T_n(x) = T(x) \qquad \text{for all } x \in E$$

if and only if the above equality holds for x in a fundamental subset of E and

$$\sup\{\|T_n\|;\ n \ge 1\} < +\infty.$$

Proof. The necessity follows from the Banach-Steinhaus theorem. To prove sufficiency, let X be a dense set in E such that the equality holds. For every $x \in E$, choose $y \in X$ such that

$$\|x-y\| < \epsilon/M$$

where $M = \sup\{\|T_n\|;\ n = 1,2,\ldots\}$. Then

$$\|T_n(x)-T(x)\| \le \|T_n(x-y)\| + \|T(x-y)\| + \|T_n(y)-T(y)\| < 3\epsilon$$

for sufficiently large n. Hence the proof is complete.

Theorem 11.8. Let T_n be a sequence of continuous linear operators from a Sargent space E into another normed linear space. If

$$\lim_{n\to\infty} T_n(x) = T(x) \quad \text{for every } x \in E,$$

then T is also continuous.

This is an immediate consequence of Theorem 11.6.

We shall give other examples of Sargent spaces. We shall construct a scale of integrals between Lebesgue's and Denjoy's. The scale is provided by the so-called α-variation. Given a function F, we define the α-variation of F for $0 < \alpha < 1$ to be

$$V_\alpha(F;[a,b]) = \sup \{ \sum_{i=1}^{n} |F(x_i)-F(x_{i-1})|^{1/\alpha} \}^\alpha$$

where the supremum is taken over all divisions $a = x_o < x_1 < \ldots < x_n = b$.

A function f is said to be D_α integrable on [a,b], or $f \in D_\alpha$, if f is Henstock integrable on [a,b] and its primitive function F has a finite α-variation. The D_α integral of f on [a,b] is defined to be the same as the Henstock integral. Note that when $\alpha = 1$ it is equivalent to the Lebesgue integral. For convenience, we may regard the D_o integral as the Denjoy integral.

Example 11.9. Let $0 < \alpha < 1$ and D_α be the space of all D_α integrable functions on [a,b] provided with the norm

$$\|f\|_\alpha = V_\alpha(F;[a,b])$$

where F is the primitive of f on [a,b]. Then we can prove that D_α is a normed linear space which is not complete but a Sargent space.

12. LINEAR FUNCTIONALS

We shall characterize continuous linear functionals on the Denjoy space. First, we prove a theorem which enables us to give an example of such a functional.

Theorem 12.1. If f is Henstock integrable on [a,b] and g is of bounded variation, then the product fg is Henstock integrable on [a,b].

Proof. Let F be the primitive of f with $F(a) = 0$. Then for every $\epsilon > 0$ there is a $\delta(\xi) > 0$ such that for any δ-fine division $D = \{[u,v];\xi\}$ we have

$$|\sum f(\xi)(v-u)-F(b)| < \epsilon.$$

Also, by Example 8.19 the Stieltjes integral of F with respect to g exists. That is, given $\epsilon > 0$ there is a $\eta > 0$ such that for any division $D = \{[u,v];\xi\}$ of [a,b] with $\xi \in [u,v] \subset (\xi-\eta,\xi+\eta)$ we have

$$|\sum F(\xi)\{g(v)-g(u)\} - \int_a^b F(x)dg(x)| < \epsilon.$$

Now write

$$A = F(b)g(b) - \int_a^b F(x)dg(x).$$

We shall prove that fg is Henstock integrable to A on [a,b].

Assume that $0 < \delta(\xi) \le \eta$. Further assume that $0 < \delta(\xi) < \xi - a$ and $0 < \delta(\xi) < b - \xi$ when $\xi \in (a,b)$. Take any δ-fine division D given by

$$a = x_o < x_1 < \ldots < x_n = b \quad \text{and} \quad \{\xi_1, \xi_2, \ldots, \xi_n\}.$$

Then we always have $\xi_1 = a$ and $\xi_n = b$. Consequently,

$$\left|\sum_{k=1}^{n} f(\xi_k)g(\xi_k)(x_k - x_{k-1}) - A\right|$$

$$= \left|\sum_{k=1}^{n-1} \{\sum_{i=1}^{k} f(\xi_i)(x_i - x_{i-1})\}\{g(\xi_k) - g(\xi_{k+1})\} + \{\sum_{i=1}^{n} f(\xi_i)(x_i - x_{i-1})\}g(\xi_n) - A\right|$$

$$\le \left|\sum_{k=1}^{n-1} \{\sum_{i=1}^{k} f(\xi_i)(x_i - x_{i-1}) - F(x_k)\}\{g(\xi_k) - g(\xi_{k+1})\}\right|$$

$$+ \left|\sum_{k=1}^{n-1} F(x_k)\{g(\xi_{k+1}) - g(\xi_k)\} - \int_a^b F(x)dg(x)\right|$$

$$+ \left|\sum_{i=1}^{n} f(\xi_i)(x_i - x_{i-1}) - F(b)\right| \, |g(b)|$$

$$< 2\epsilon V(g) + \epsilon + \epsilon |g(b)|$$

where V(g) denotes the total variation of g on [a,b]. Therefore fg is Henstock integrable on [a,b].

Corollary 12.2. If f is Henstock integrable on [a,b] with the primitive F and g of bounded variation, then

$$\int_a^b f(x)g(x)dx = F(b)g(b) - F(a)g(a) - \int_a^b F(x)dg(x)$$

where the integral on the right is in the Stieltjes sense.

This is the formula of integration by parts. Using the formula, we can show that the norm $\|f\|$ in the Denjoy space as given in Example 11.1 is equivalent to the following

$$\|f\|^* = \sup \left|\int_a^b f(x)g(x)dx\right|$$

where the supremum is taken over all functions g of bounded variation such that $|g(x)| \le 1$ for all $x \in [a,b]$ and the total variation $V(g;[a,b]) \le 1$. Indeed, we have

$$\|f\| \le \|f\|^* \le 2\|f\|.$$

Theorem 12.3. If g is of bounded variation on [a,b] then

$$T(f) = \int_a^b f(x)g(x)dx$$

defines a continuous linear functional on the Denjoy space with

$$\|T\| \le |g(b)| + V(g;[a,b]).$$

Proof. Using integration by parts, we have

$$|T(f)| \le \|f\| \{|g(b)| + V(g;[a,b])\}$$

where $\|f\|$ denotes the norm in the Denjoy space. Hence T is bounded and therefore continuous with $\|T\|$ satisfying the required inequality.

The above provides an example of a functional which is linear and continuous on the Denjoy space. Now we may extend the controlled convergence theorem slightly.

Theorem 12.4. If f_n is control-convergent to f on [a,b] and g is of bounded variation, then fg is Henstock integrable on [a,b] and

$$\int_a^b f_n(x)g(x)dx \to \int_a^b f(x)g(x)dx \text{ as } n \to \infty.$$

Proof. The controlled convergence theorem shows that f is Henstock integrable on [a,b]. It follows from Theorem 12.1 that fg is Henstock integrable on [a,b]. Let F_n be the primitive of f_n and F that of f. Apply integration by parts and obtain

$$\int_a^b \{f_n(x) - f(x)\}g(x)dx = \{F_n(b) - F(b)\}g(b) - \int_a^b \{F_n(x)-F(x)\}dg(x).$$

Since the right side tends to zero by the uniform convergence theorem for the Stieltjes integral, so does the left side.

Next, we prove a lemma due to Riesz.

Definition 12.5. A function F is said to satisfy the Lipschitz condition if there is a $M > 0$ such that

$$|F(y)-F(x)| \le M|y-x| \quad \text{for all } x, y.$$

A function F is said to be of bounded slope variation on [a,b] if

$$\sum_{i=0}^{n-2} \left| \frac{F(x_{i+2}) - F(x_{i+1})}{x_{i+2} - x_{i+1}} - \frac{F(x_{i+1}) - F(x_i)}{x_{i+1} - x_i} \right|$$

is bounded for all divisions $a = x_o < x_1 < \ldots < x_n = b$.

Lemma 12.6. A function F is the primitive of a function of bounded variation on [a,b] if and only if F satisfies the Lipschitz condition and is of bounded slope variation on [a,b].

Proof. Suppose f is of bounded variation on [a,b] with the primitive F. Then f is bounded and F satisfies the Lipschitz condition. Further, the sum as given in Definition 12.5 is less than

$$\sum_{i=0}^{n-2} \omega(f;[x_i,x_{i+2}]) \le 2\, V(f;[a,b]).$$

Hence it is bounded by 2 V(f;[a,b]) for all divisions of [a,b].

Conversely, suppose F satisfies the conditions. Let $[a_i,b_i]$, $i = 1,2,\ldots, n$, be a finite sequence of non-overlapping intervals, and write $s_o = 0$ and $s_k = (b_1-a_1) + \ldots + (b_k-a_k)$ for $k = 1,2,\ldots, n$. Then

$$\sum_{i=1}^{n} |F(b_i) - F(a_i)| = \sum_{i=1}^{n-1} \left\{ \left|\frac{F(b_i)-F(a_i)}{b_i-a_i}\right| - \left|\frac{F(b_{i+1})-F(a_{i+1})}{b_{i+1} - a_{i+1}}\right| \right\} s_j + \left|\frac{F(b_n)-F(a_n)}{b_n-a_n}\right| s_n$$

$$\le Ms_n$$

for some $M > 0$. Hence F is absolutely continuous on [a,b].

Now let $f(x) = F'(x)$ almost everywhere in [a,b]. Put $F_n(x) = F(x)$ when $x = a + k(b - a)/2^n$, $k = 0,1,2,\ldots,2^n$, and linearly elsewhere. Then $f_n(x) = F_n'(x)$ converges almost everywhere to $f(x)$ as $n \to \infty$. At each point x for which $F'(x)$ exists we have

$$\frac{F(v) - F(u)}{v - u} \longrightarrow f(x) \quad \text{as } u,v \longrightarrow x$$

where u and v are the dyadic points above and $u \le x \le v$. It follows from the condition of bounded slope variation that

$$\sum_{i=1}^{n} |f(x_i)-f(x_{i-1})| \le M$$

where $a \le x_o < x_1 < \ldots x_n \le b$ and x_i are taken over all the points x at which $F'(x)$ exists. Hence f is equal to a function of bounded variation almost everywhere. However we may take f itself to be of bounded variation, and F is the primitive of f.

Let $V(g)$ denote the total variation of g on $[a,b]$ and

$$\|g\|_{EBV} = \inf V(g_1)$$

where the infimum is over all g_1 of bounded variation and such that $g_1(x) = g(x)$ almost everywhere in $[a,b]$.

Theorem 12.7. If T is a continuous linear functional on the Denjoy space E, then

$$T(f) = \int_a^b f(x)g(x)dx$$

for all $f \in E$ and for some g of bounded variation on $[a,b]$. Furthermore,

$$\|g\|_{EBV} \leq 2\|T\|.$$

Proof. Put $G(t) = T(\chi_{[a,t]})$ where $\chi_{[a,t]}$ denotes the characteristic function of $[a,t]$. Take a division $a = x_o < x_1 < \ldots < x_n = b$. Then by the linearity of T we obtain

$$\sum_{i=0}^{n-2} \left| \frac{G(x_{i+2})-G(x_{i+1})}{x_{i+2}-x_{i+1}} - \frac{G(x_{i+1})-G(x_i)}{x_{i+1}-x_i} \right| = \sum_{i=0}^{n-2} |T(\varphi_i)|$$

where

$$\varphi_i = \frac{1}{x_{i+2}-x_{i+1}} \chi_{(x_{i+1},x_{i+2}]} - \frac{1}{x_{i+1}-x_i} \chi_{(x_i,x_{i+1}]}.$$

Further by the boundedness of T we obtain

$$\sum_{i=0}^{n-2} |T(\varphi_i)| = T(\sum_{i=0}^{n-2} \epsilon_i\varphi_i) \leq \|T\| \|\sum_{i=0}^{n-2} \epsilon_i\varphi_i\| \leq 2\|T\|$$

where ϵ_i denotes $+1$ or -1 as the case may be. That is, G is of bounded slope variation. Similarly, we can show that G also satisfies the Lipschitz condition. It follows from Lemma 12.6 that G is the primitive of a function g which is of bounded variation on $[a,b]$. Therefore the representation holds true for step functions. Note that taking $x_o, x_1, \ldots, x_n$ to be the dyadic points as in the proof of Lemma 12.6 we can show that $\|g\|_{EBV} \leq 2\|T\|$.

Let f be Henstock integrable on $[a,b]$. In view of the

Riesz-type definition (Theorem 10.2), there is a sequence of step functions φ_n control-converging to f. Applying Theorem 12.4, we have

$$T(f) = \lim_{n\to\infty} T(\varphi_n)$$

$$= \lim_{n\to\infty} \int_a^b \varphi_n(x)g(x)dx$$

$$= \int_a^b f(x)g(x)dx$$

Hence the proof is complete.

Let g be an almost everywhere bounded function on [a,b] and define

$$\|g\|_\infty = \inf\{M; |g(x)| \le M \quad \text{almost everywhere}\}.$$

Theorem 12.8. If fg is Henstock integrable on [a,b] for every absolutely Henstock integrable function f on [a,b], then g is almost everywhere bounded on [a,b].

Proof. Let L_1 denote the space of all absolutely Henstock integrable functions on [a,b] provided with the norm

$$\|f\| = \int_a^b |f(x)|dx.$$

Put $g_n(x) = g(x)$ when $|g(x)| \le n$ and zero otherwise, and define

$$T_n(f) = \int_a^b f(x)g_n(x)dx.$$

Then each T_n is a continuous linear functional on L_1 and, furthermore,

$$\|T_n\| = \|g_n\|_\infty.$$

Indeed, if we take f(x) = 1 when $u \le x \le v$ and 0 otherwise then

$$\left|\frac{1}{v-u}\int_u^v g_n\right| = \left|\int_a^b fg_n\right| \le \|T_n\| \; \|f\| = \|T_n\|.$$

Since the primitive of g_n is differentiable almost everywhere, we obtain $\|g_n\|_\infty \le \|T_n\|$. The reserve inequality is easy.

By the dominated convergence theorem, for $f \in L_1$ we have

$$\lim_{n\to\infty} T_n(f) = \int_a^b f(x)g(x)dx.$$

Therefore apply the Banach-Steinhaus theorem and we obtain that g is bounded almost everywhere.

Theorem 12.9. If fg is Henstock integrable on [a,b] for every Henstock integrable function f, then g is almost everywhere equal to a function of bounded variation on [a,b].

Proof. Since fg is Henstock integrable on [a,b] for every absolutely Henstock integrable f, then by Theorem 12.8, g is at least bounded almost everywhere.

Now suppose g is not almost everywhere equal to a function of bounded variation. That is, for any function $g_1(x) = g(x)$ almost everywhere, g_1 is not of bounded variation. Then there is a point $c \in [a,b]$ such that g_1 is not of bounded variation on any open interval containing c. Thus there exists an increasing (or decreasing) sequence $\{x_n\}$ with the limit c such that

$$\sum_{n=1}^{\infty} (M_n - m_n) = +\infty$$

where $M_n = \sup\{g_1(x); x_n \le x \le x_{n+1}\}$ and $m_n = \inf\{g_1(x); x_n \le x \le x_{n+1}\}$.

Corresponding to each n, there exist distinct measurable subsets X_n and Y_n of $[x_n, x_{n+1}]$ such that

$$g_1(x) \ge \frac{3}{4}M_n + \frac{1}{4}m \quad \text{when } x \in X_n,$$

$$g_1(x) \le \frac{1}{4}M_n + \frac{3}{4}m_n \quad \text{when } x \in Y_n,$$

and the measures of X_n and Y_n are positive and equal, say, δ_n. Now write

$$p_n = \{\delta_n \sum_{i=1}^{n} (M_i - m_i)\}^{-1}$$

and define $f(x) = p_n$ when $x \in X_n$, $f(x) = -p_n$ when $x \in Y_n$ for $n = 1,2,\ldots$ and 0 elsewhere. Since f is Henstock integrable on each [a,u] with $x_n \le u < x_{n+1}$ and

$$\left|\int_a^u f(x)dx\right| \le 2p_n\delta_n$$

which tends to zero as $n \to \infty$, hence f is Henstock integrable on [a,b].

On the other hand,

$$\int_{x_n}^{x_{n+1}} f(x)g_1(x)dx \geq p_n\delta_n(\tfrac{3}{4}M_n + \tfrac{1}{4}m_n) - p_n\delta_n(\tfrac{1}{4}M_n + \tfrac{3}{4}m_n)$$

$$\geq \frac{1}{2}(M_n - m_n)/\sum_{i=1}^{n}(M_i - m_i).$$

Summing both sides over n = 1,2,... and making use of the fact that if $\sum_{n=1}^{\infty}(M_n - m_n) = +\infty$ then

$$\sum_{n=1}^{\infty}\{(M_n - m_n)/\sum_{i=1}^{n}(M_i - m_i)\} = +\infty,$$

we find that fg_1 is not Henstock integrable on [a,c]. It leads to a contradiction. Hence g is almost everywhere a function of bounded variation on [a,b].

For completeness, we verify the above result on divergent series here. Write $a_n = M_n - m_n > 0$ and s_n its partial sum. We observe that

$$\frac{a_{n+1}}{s_{n+1}} + \ldots + \frac{a_{n+k}}{s_{n+k}} \geq \frac{a_{n+1} + \ldots + a_{n+k}}{s_{n+k}} = 1 - \frac{s_n}{s_{n+k}}.$$

Since $s_n \to \infty$ as $n \to \infty$, we can choose k(n) such that $s_n/s_{n+k(n)} < 1/2$ and therefore

$$\sum_{i=n}^{n+k(n)} \frac{a_i}{s_i} > \frac{1}{2}.$$

Hence the resulting series $\sum_{n=1}^{\infty} a_n/s_n$ diverges. This is the well-known theorem of Abel and Dini (1867).

Theorem 12.10. The following two conditions are equivalent:

(i) For every Henstock integrable function f on [a,b] there is a constant M(f) such that for all n

$$\left|\int_a^b f(x)g_n(x)dx\right| \leq M(f);$$

(ii) $\sup\{\|g_n\|_\infty + \|g_n\|_{EBV};\ n\geq 1\} < +\infty.$

Proof. Obviously, (ii) implies (i). Assume (i). It follows from Theorem 12.9 that each g_n is almost everywhere equal to a function of bounded variation on [a,b]. Define

$$T_n(f) = \int_a^b f(x)g_n(x)dx.$$

Then T_n is a continuous linear functional on the Denjoy space as well as on the space of all absolutely Henstock integrable functions on $[a,b]$. Apply the Banach-Steinhaus theorem for Sargent spaces (Theorem 11.6) and we obtain from the former that

$$\sup\{\|g_n\|_{EBV};\ n\geq 1\} < +\infty,$$

and from the latter that $\sup\{\|g_n\|_\infty;\ n\geq 1\} < +\infty$.

Theorem 12.11. Let E be the Denjoy space. In order that

$$\int_a^b f(x)g_n(x)dx \longrightarrow \int_a^b f(x)dx \quad \text{as } n \to \infty$$

whenever $f \in E$, it is necessary and sufficient that

(i) g_n is almost everywhere of bounded variation on $[a,b]$ for each n and

$$\sup\ \{\|g_n\|_\infty + \|g_n\|_{EBV};\ n\geq 1\} < +\ \infty;$$

(ii) $\int_c^d g_n(x)dx \to d - c$ as $n \to \infty$ whenever $a \leq c < d \leq b$.

Proof. Condition (i) follows from Theorem 12.10. Condition (ii) is trivial. The sufficiency follows from Theorem 11.7 since the set of step functions is dense in the Denjoy space E and therefore the set of characteristic functions of subintervals $[c,d]$ is fundamental in E.

We end the section by giving an alternative proof to the classical Riesz representation theorem without resorting to the Hahn-Banach extension theorem. We recall that the Stieltjes integral was defined immediately after Definition 8.18.

Theorem 12.12. Let C be the space of continuous functions on $[a,b]$ with the uniform norm. If T is a continuous linear functional on C, then

$$T(f) = \int_a^b f(x)dg(x)$$

for all $f \in C$ and for some function g of bounded variation on $[a,b]$ where the integral is in the Stieltjes Sense.

Proof. Suppose F is continuous and ACG* on $[a,b]$. Then F is the primitive of a Henstock integrable function f on $[a,b]$. Define

$T^*(f) = T(F)$ for all such f and applying Theorem 12.7 we have

$$T(F) = T^*(f) = \int_a^b f(x)g(x)dx$$

for all such f and for some function g of bounded variation on [a,b]. Using integration by parts, we write

$$T(F) = F(b)g(b) - \int_a^b F(x)dg(x) = \int_a^b F(x)dg_1(x)$$

where $g_1(x) = -g(x)$ when $x \in [a,b)$ and $g_1(b) = 0$.

Since every continuous function F is the limit of a uniformly convergent sequence of continuous functions $\{F_n\}$ which are also ACG*, apply the uniform convergence theorem for the Stieltjes integral and we obtain

$$\begin{aligned} T(F) &= \lim_{n\to\infty} T(F_n) \\ &= \lim_{n\to\infty} \int_a^b F_n(x)dg_1(x) \\ &= \int_a^b F(x)dg_1(x). \end{aligned}$$

Hence T can be represented by means of the Stieltjes integral.

13. A NONLINEAR INTEGRAL

In order to prove Riesz representation theorems for nonlinear functionals, we develop a nonlinear integral. It is obtained by making a slight change in the definition of the Henstock integral.

Definition 13.1. Let $\phi = \phi(s,I)$ be defined for s being real and I an interval. A function f is said to be ϕ-integrable on [a,b] if there is a number A such that for every $\epsilon > 0$ there is a $\delta(\xi) > 0$ such that for any δ-fine division $D = \{[u,v];\xi\}$ we have

$$|\sum\phi(f(\xi),[u,v]) - A| < \epsilon.$$

The ϕ-integral of f on [a,b] is A, which is uniquely determined, and we write

$$\int_a^b \phi(f(\cdot),\cdot) = A$$

If we put $\phi(s,[u,v]) = s(v-u)$, then the above reduces to the

Henstock integral. In general, $\phi(s,I)$ represents the measure of a single-step function having value s on I and zero elsewhere. Since $\phi(s,I)$ is not necessarily linear in s, the integral defined above is a nonlinear integral.

In order that the integral has some reasonable properties, we assume in what follows and throughout this section that the measure ϕ satisfies the following conditions:

(N1) $\phi(0,I) = 0$;

(N2) $\phi(.,I)$ is continuous;

(N3) $\phi(s,I_1 \cup I_2) = \phi(s,I_1) + \phi(s,I_2)$ whenever I_1 and I_2 are adjacent and disjoint;

(N4) given $M > 0$, for every $\epsilon > 0$ there exists a $\eta > 0$ such that

$$\left|\sum_{i=1}^{n} \phi(s_i,I_i) - \sum_{i=1}^{n} \phi(t_i,I_i)\right| < \epsilon$$

whenever $|s_i-t_i| < \eta$, $|s_i| \le M$ and $|t_i| \le M$ for $i = 1, 2, \ldots, n$, and $I_1, I_2, \ldots, I_n$ are pairwise disjoint intervals in $[a,b]$;

(N5) given $M > 0$, for every $\epsilon > 0$ there is a $\eta > 0$ such that

$$\left|\sum_{i=1}^{n} \phi(s_i,I_i)\right| < \epsilon$$

whenever $I_1, I_2, \ldots, I_n$ are pairwise disjoint intervals with the total length less than η and $|s_i| \le M$ for $i = 1,2,\ldots,n$.

Obviously, (N4) implies (N2). Also we assume throughout that

$$\phi(s,[u,v]) = \phi(s,[u,v)) = \phi(s,(u,v)) = \phi(s,(u,v]).$$

Example 13.2. Let f be a step function, or more precisely, let $I_1,I_2,\ldots,I_n$ be pairwise disjoint subintervals of $[a,b]$ and $f(x) = s_i$ when $x \in I_i$ for $i = 1, 2, \ldots, n$. Then making use of (N5) we obtain that f is ϕ-integrable on $[a,b]$ and

$$\int_a^b \phi(f(\cdot),\cdot) = \sum_{i=1}^{n} \phi(s_i,I_i).$$

Example 13.3. Every continuous function f is ϕ-integrable on $[a,b]$. The proof is to find a suitable $\delta(\xi) > 0$ for given $\epsilon > 0$. Since f is continuous on $[a,b]$ it is uniformly continuous there. That is, for every $\eta > 0$ there exists a $\delta > 0$ such that

$$|f(x) - f(y)| < \eta \qquad \text{whenever } |x-y| < \delta.$$

For any division $D = \{[u,v];\xi\}$ we write

$$\sigma(f,D) = \Sigma\phi(f(\xi),[u,v]).$$

Given $\epsilon > 0$, choose a $\eta > 0$ as in (N4) and put $\delta(\xi) = \delta/2$ for all ξ. Then for any two δ-fine divisions D_1 and D_2 there exists another δ-fine division D_3 which is finer than both D_1 and D_2 and by using (N4) we have

$$|\sigma(f,D_1)-\sigma(f,D_2)| \le |\sigma(f,D_1)-\sigma(f,D_3)| + |\sigma(f,D_3)-\sigma(f,D_2)| < 2\epsilon.$$

Hence the Cauchy condition holds and f is ϕ-integrable on [a,b].

Example 13.4. If $f(x) = 0$ almost everywhere, then f is ϕ-integrable on [a,b]. The proof is similar to that of Theorem 3.5 with the use of (N5).

We remark that (N1) ensures that the ϕ-integral of the zero function is zero. Condition (N2) will be required later in the proof of convergence theorems. In order to obtain

$$\int_a^c \phi(f(\cdot),\cdot) + \int_c^b \phi(f(\cdot),\cdot) = \int_a^b \phi(f(\cdot),\cdot),$$

we need (N3). Condition (N4) is used in Example 13.3 and (N5) in Example 13.2 and again in Example 13.4. Note that all the conditions are trivially satisfied when $\phi(s,[u,v]) = s(v-u)$.

It is easy to verify that the simple properties given in Section 3 are valid for the ϕ-integral. In particular, Henstock's lemma holds.

Theorem 13.5. If f is ϕ-integrable on [a,b] then for every $\epsilon > 0$ there is a $\delta(\xi) > 0$ such that for any δ-fine division $D = \{[u,v];\xi\}$ of [a,b] we have

$$\sum|F_\phi(u,v)-\phi(f(\xi),[u,v])| < \epsilon$$

where

$$F_\phi(u,v) = \int_u^v \phi(f(\cdot),\cdot).$$

As usual, we call F_ϕ the ϕ-primitive of f on [a,b]. Here $F_\phi(u,v) = F_\phi(v)-F_\phi(u)$.

Theorem 13.6. Let f_n be ϕ-integrable on [a,b] for $n = 1,2,\ldots$ and satisfy the following conditions:

(i) $f_n(x) \to f(x)$ almost everywhere in [a,b] as $n\to\infty$;

(ii) the ϕ-primitives $F_{\phi,n}$ of f_n are ACG^* uniformly in n;

(iii) $F_{\phi,n}(x)$ converges pointwise to a continuous function $F_\phi(x)$ on [a,b],

then f is ϕ-integrable on [a,b] with ϕ-primitive F_ϕ.

Proof. The proof is the same as that of the Henstock integral. For completeness, we sketch as follows. First, we observe that f is ϕ-integrable on [a,b] if and only if there exists a continuous function F_ϕ which is ACG^* and such that for almost all $\xi \in [a,b]$ with $u \le \xi \le v$ and $u \ne v$ we have

$$\{F_\phi(u,v)-\phi(f(\xi),[u,v])\}/(v-u) \to 0 \quad \text{as } (v-u) \to 0.$$

The sufficiency follows from the proof of Theorem 6.12, and the necessity that of Theorem 6.13. Since ϕ satifies (N5), the fact that

$$\sum|f(\xi)(v-u)| < \epsilon$$

when $|f(\xi)| \le M$ and $\sum|v-u| < \eta$ in the original proof, is now replaced by

$$\sum|\phi(f(\xi),[u,v])| < \epsilon$$

using (N5). The boundedness of $\sum|\phi(f(\xi),[u,v])|$ also follows from (N5), which is used in the proof that F_ϕ is VBG^* (Lemma 6.17).

Next, we apply the linearization process as in the proof of Theorem 7.6. The ordinary derivative in the original proof is now replaced by the derivative with respect to ϕ as defined above. Hence we can apply the corresponding theorem of Theorem 7.1 for the ϕ-integral and prove that f is ϕ-integrable on [a,b] with ϕ-primitive F_ϕ.

We remark that all the conditions (N1) to (N5) are used in the proof of the above theorem. Furthermore, if condition (ii) in the above theorem is replaced by: $F_{\phi,n}$ are uniformly absolutely continuous on [a,b], then, as shown in Theorem 7.1, condition (iii) can now be deduced and is therefore not required.

Definition 13.7. Let X be a function space. A functional T defined on X is said to be orthogonally additive if

$$T(f + g) = T(f) + T(g)$$

whenever f, $g \in X$ and have disjoint supports.

We recall that two functions f and g have disjoint supports if

$f(x)g(x) = 0$ for all x, or $f(x)g(x) = 0$ almost everywhere in the case when f and g equal almost everywhere are regarded as identical. It is an easy consequence of (N1) that

$$T(f) = \int_a^b \phi(f(\cdot),\cdot),$$

as a functional defined on the space of continuous functions, is orthogonally additive. Furthermore, any linear functional is orthogonally additive.

Definition 13.8. Let E denote the Denjoy space of all Henstock integrable functions on [a,b]. A functional T defined on E is said to be control-continuous if

$$\lim_{n\to\infty} T(f_n) = T(f)$$

whenever f_n is control-convergent to f on [a,b].

Example 13.9. Let g be of bounded variation on [a,b]. Then by Theorem 12.4

$$T(f) = \int_a^b f(x)g(x)dx$$

defines a control continuous functional on the Denjoy space E. In fact, any norm continuous functional on E is also control-continuous. We shall prove later on (see the remark after Theorem 15.12) that the two are equivalent for linear functionals on E.

Suppose the following integral

$$\int_a^b \phi(f(\cdot),\cdot)$$

exists for all Henstock integrable functions f, and suppose f_n is control-convergent to f on [a,b]. In order to apply Theorem 13.6, we require the following property.

Definition 13.10. Let f_n be Henstock integrable on [a,b] with primitive F_n and ϕ-integrable on [a,b] with ϕ-primitive $F_{\phi,n}$ for $n = 1,2,\ldots$. A measure ϕ as defined in Definiton 13.1 is said to be control-invariant if the following conditions are satisfied:

(i) if F_n are ACG^* uniformly in n, then so are $F_{\phi,n}$;

(ii) if F_n are equicontinuous then so are $F_{\phi,n}$.

For convenience, we call property (i) the $UACG^*$ invariance, and property (ii) the equicontinuity invariance.

Example 13.11. Let ϕ be a control-invariant measure. Then

$$T(f) = \int_a^b \phi(f(\cdot),\cdot)$$

defines a control-continuous functional on the Denjoy space E. This is so by means of Theorem 13.6.

14. A THEOREM OF DREWNOWSKI AND ORLICZ

We shall represent orthogonally additive functionals by means of the nonlinear integral as defined in Section 13. The difficult step is to prove the necessity of (N4). This is essentially a theorem of Drewnowski and Orlicz (Theorem 14.8). In the next section, we shall prove a representation theorem for control-continuous orthogonally additive functionals defined on the Denjoy space (Theorem 15.10). In preparation, we prove here a result for a smaller function space, namely, L_∞.

As usual, L_∞ denotes the space of all essentially bounded measurable functions on [a,b]. That is, $f \in L_\infty$ if and only if f is measurable and bounded almost everywhere in [a,b]. Two functions are regarded as identical if they are equal almost everywhere. The norm in L_∞ is given by

$$\|f\|_\infty = \inf\{M;\ |f(x)| \le M \text{ almost everywhere in } [a,b]\}.$$

Definition 14.1. A sequence of measurable functions $\{f_n\}$ is said to be boundedly convergent to f if $f_n(x) \to f(x)$ almost everywhere as $n \to \infty$ and $\{f_n\}$ is uniformly bounded almost everywhere by a constant. A functional T defined on L_∞ is said to be boundedly continuous if $T(f_n) \to T(f)$ as $n \to \infty$ whenever $\{f_n\}$ is boundedly convergent to f.

Theorem 14.2. Let ϕ satisfy (N1)–(N5). If f is ϕ-integrable on [a,b] for all $f \in L_\infty$ then

$$T(f) = \int_a^b \phi(f(\cdot),\cdot)$$

defines a boundedly continuous and orthogonally additive functional on L_∞.

Proof. First, we show that if F_ϕ is the ϕ-primitive of $f \in L_\infty$ then F_ϕ is absolutely continuous. In view of (N5), for every $M > 0$ and $\epsilon > 0$ there is an $\eta > 0$ such that

$$\left|\sum_{i=1}^{n} \phi(s_i, I_i)\right| < \epsilon$$

whenever I_i, $i = 1,2,\ldots,n$, are pairwise disjoint intervals with the total length less than η and $|s_i| \le M$ for all i. Let $f \in L_\infty$ and $|f(x)| \le M$ for all x. Since f is ϕ-integrable, there is a $\delta(\xi) > 0$ such that for any δ-fine division $\{[u,v];\xi\}$ we have

$$\sum |F_\phi(u,v) - \phi(f(\xi), [u,v])| < \epsilon.$$

Then whenever $\sum_{i=1}^{n} |I_i| < \eta$ consider δ-fine division D_i of I_i and $\sum_i$ sums over D_i and we obtain

$$\begin{aligned}\sum_{i=1}^{n} |F_\phi(I_i)| &\le \epsilon + \sum_{i=1}^{n} \sum_i |\phi(f(\xi), [u,v])| \\ &< 2\epsilon.\end{aligned}$$

That is, F_ϕ is absolutely continuous on $[a,b]$.

Note that in the above proof η was chosen independent of f. Hence we have also shown that if $F_{\phi,n}$ is the ϕ-primitive of $f_n \in L_\infty$ and $|f_n(x)| \le M$ for all x and all n then $F_{\phi,n}$ are uniformly absolutely continuous.

Now suppose f_n is boundedly convergent to f. Then the ϕ-primitives $F_{\phi,n}$ of f_n are uniformly absolutely continuous and by the controlled convergence theorem for the ϕ-integral (Theorem 13.6) and the remark thereafter

$$T(f_n) \to T(f) \quad \text{as } n \to \infty.$$

Note that all the conditions (N1) to (N5) are used, for example (N4) is used in the proof of the controlled convergence theorem for the ϕ-integral. The orthogonal additivity of T follows from (N1), and the proof is complete.

In fact, the converse of Theorem 14.2 also holds. To prove the converse, we need the following series of lemmas.

Lemma 14.3. Let f_n be measurable functions dominated by another measurable function on $[a,b]$. Then there is a sequence of integers $\{m(n)\}$ such that for almost all $x \in [a,b]$

$$\limsup_{n\to\infty} f_n(x) = \lim_{n\to\infty} \max\{f_m(x);\ n \le m \le m(n)\}.$$

Proof. Let $g_{n,m}(x) = \sup\{f_k(x);\ n\le k\le m\}$. Denote by $g_n(x)$ the limit of $g_{n,m}(x)$ as $m \to \infty$ and by $f(x)$ the limit of $g_n(x)$ as $n \to \infty$. Note that $f(x)$ is the upper limit of $f_n(x)$. By Egoroff's theorem (Lemma 7.2), there is an open set G_n with $|G_n| < 2^{-n}$ such that $g_{n,m}(x)$ converges uniformly to $g_n(x)$ on $[a,b] - G_n$ as $m \to \infty$.

Now choose $m(n)$ such that

$$|g_{n,m}(x)-g_n(x)| < 1/n$$

for $m \ge m(n)$ and all $x \in [a,b]-G_n$. Write

$$S = \bigcap_{n=1}^{\infty} \bigcup_{k=n}^{\infty} G_k.$$

Then $|S| = 0$ and for $x \in [a,b]-S$ it follows from

$$|g_{n,m(n)}(x)-f(x)| \le |g_{n,m(n)}(x)-g_n(x)| + |g_n(x)-f(x)|$$

that $g_{n,m(n)}(x) \to f(x)$ as $n \to \infty$. Hence the required result holds.

Lemma 14.4. If T is boundedly continuous and orthogonally additive on L_∞, then there is a function k such that $k(0,x) = 0$ for almost all x, $k(s,\cdot)$ is absolutely Henstock integrable on $[a,b]$ for each real value s and for any simple function f

$$T(f) = \int_a^b k(f(x),x)dx.$$

Proof. Fix s and write $F(x) = T(s\chi_{[a,x]})$ where $\chi_{[a,x]}$ is the characteristic function of $[a,x]$. Since T is boundedly continuous, F is absolutely continuous on $[a,b]$. It follows from Theorem 5.8 that $k(s,x) = F'(x)$ exists almost everywhere and consequently

$$T(s\chi_E) = \int_E k(s,x)dx$$

for any measurable set E in $[a,b]$. Since T is orthogonally additive, we have $k(0,x) = 0$ for almost all x and hence the required equality holds.

We remark that actually the above equality can be proved to hold true for bounded measurable functions (Theorem 14.10). The function k is called a kernel function.

Lemma 14.5. Let k be a kernel function as given in Lemma 14.4, and let f_i and g_i be measurable functions satisfying $|f_i(x)| \le M$ and $|g_i(x)| \le M$ for $i = 1,2,\ldots,n$ and all $x \in [a,b]$. If

$$u(x) = \max\{f_i(x);\ 1 \le i \le n\}$$

then there is a measurable function v bounded by M such that for almost all $x \in [a,b]$

$$|u(x)-v(x)| \le \max\{|f_i(x)-g_i(x)|;\ 1 \le i \le n\},$$

$$\min\{|k(f_i(x),x)-k(g_i(x),x)|;\ 1 \le i \le n\} \le |k(u(x),x)-k(v(x),x)|$$
$$\le \max|k(f_i(x),x)-k(g_i(x),x)|;\ 1 \le i \le n\}.$$

Proof. In view of the definition of u, we can find disjoint measurable sets $E_1, E_2, \ldots, E_n$ such that

$$u(x) = \sum_{i=1}^{n} f_i(x)\chi_{E(i)}(x)$$

where E(i) denotes E_i. Then define

$$v(x) = \sum_{i=1}^{n} g_i(x)\chi_{E(i)}(x).$$

It is easy to verify that the conditions in the lemma are satisfied.

Lemma 14.6. Let k be a kernel function as given in Lemma 14.4. If f_i are simple functions satisfying $|f_i(x)| \le M$ for $i = 1,2,\ldots$ and all $x \in [a,b]$, then the sequence $\{k(f_i(x),x)\}$ is dominated by an absolutely Henstock integrable function on [a,b].

Proof. Fix n and let E_i denote the set of all x such that

$$|k(f_i(x),x)| = \max\{|k(f_k(x),x)|;\ 1 \le k \le n\}$$

Put $E_1^* = E_1$ and $E_i^* = E_i-(E_1 \cup\ldots\cup E_{i-1})$ for $i = 2,\ldots,n$. Then define

$$u_n = \sum_{i=1}^{n} f_i\chi_{E_i^*}$$

where χ_E denotes the characteristic function of E. Obviously, $|u_n(x)| \le M$ for all x and

$$|k(u_n(x),x)| = \max\{|k(f_i(x),x)|;\ 1\le i\le n\}.$$

It remains to show that the sequence $\{k(u_n(x),x)\}$ is dominated.

In view of Lemma 14.3, we can find $v_n(x) = \max\{u_n(x),\dots, u_{m(n)}(x)\}$ such that for almost all x

$$\limsup_{n\to\infty} u_n(x) = \lim_{n\to\infty} v_n(x).$$

Then, by Lemma 14.5 with u replaced by v_n and both g_i and v being zero, we have

$$|k(u_n(x),x)| \le |k(v_n(x),x)| \le |k(u_{m(n)}(x),x)|.$$

That is, for almost all x

$$\lim_{n\to\infty} |k(u_n(x),x)| = \lim_{n\to\infty}|k(v_n(x),x)|.$$

Since $\{v_n\}$ is a sequence of simple functions boundedly convergent to a limit function, say u, by Lemma 14.4 we obtain

$$\begin{aligned}\lim_{n\to\infty}\int_a^b k(v_n(x),x)dx &= \lim_{n\to\infty} T(v_n)\\ &= T(u).\end{aligned}$$

Indeed, the above holds for any boundedly convergent sequence of simple functions $\{v_n\}$. Now let

$$E^+_{nm} = \{x; k(v_n(x),x)-k(v_m(x),x) \ge 0\},\ E^-_{nm} = [a,b]-E^+_{nm},$$

$$E^+_m = \{x; k(u(x),x)-k(v_m(x),x) \ge 0\},\ E^-_m = [a,b]-E^+_m.$$

Then we have

$$\begin{aligned}&\int_a^b |k(v_n(x),x)-k(v_m(x),x)|\\ &= T(v_n\chi_{E^+_{nm}})-T(v_m\chi_{E^+_{nm}})-T(v_n x_{E^-_{nm}})+T(v_m\chi_{E^-_{nm}})\\ &= T(v_n\chi_{E^+_{nm}} + v_m\chi_{E^-_{nm}})-T(v_m\chi_{E^+_{nm}} + v_n\chi_{E^-_{nm}})\end{aligned}$$

As $n\to\infty$ we obtain

$$v_n\chi_{E^+_{nm}} + v_m\chi_{E^-_{nm}} \to u\chi_{E^+_m} + v_m\chi_{E^-_m} \quad \text{almost everywhere,}$$

$$v_m\chi_{E^+_{nm}} + v_n\chi_{E^-_{nm}} \to v_m\chi_{E^+_m} + u\chi_{E^-_m} \quad \text{almost everywhere,}$$

and furthermore as $m \to \infty$ we obtain

$$u\chi_{E_m^+} + v_m\chi_{E_m^-} \to u \quad \text{almost everywhere,}$$

$$v_m\chi_{E_m^+} + u\chi_{E_m^-} \to u \quad \text{almost everywhere.}$$

Consequently, we prove that

$$\lim_{m\to\infty} \lim_{n\to\infty} \int_a^b |k(v_n(x),x)-k(v_m(x),x)|dx$$

$$= \lim_{m\to\infty} T(u\chi_{E_m^+} + v_m\chi_{E_m^-}) - \lim_{m\to\infty} T(v_m\chi_{E_m^+} + u\chi_{E_m^-})$$

$$= T(u)-T(u) = 0$$

Then there is a subsequence $\{k(v_{n(i)}(x),x)\}$ which is dominated by an absolutely Henstock integrable function on $[a,b]$. Hence $\{k(u_n(x)x)\}$ is dominated and consequently so is $\{k(f_i(x),x)\}$.

Lemma 14.7. Let k be a kernel function as given in Lemma 14.4, and let f_i and g_i be simple functions satisfying $|f_i(x)| \le M$ and $|g_i(x)| \le M$ for $i = 1,2,\ldots$ and all $x \in [a,b]$. If $\|f_i-g_i\|_\infty \to 0$ as $i \to \infty$ then for almost all $x \in [a,b]$

$$k(f_i(x),x)-k(g_i(x),x) \to 0 \qquad \text{as } i \to \infty.$$

Proof. Suppose there is $\epsilon > 0$ such that $|E| > 0$ where

$$E = \{x;\ \limsup_{n\to\infty}|k(f_n(x),x)-k(g_n(x),x)| > \epsilon\}.$$

It follows from Lemma 14.6 that $\{k(f_n(x),x)-k(g_n(x),x)\}$ is dominated by an absolutely Henstock integrable function on $[a,b]$. In view of Lemma 14.3, we can find $m(n)$ such that

$$\limsup_{n\to\infty}|k(f_n(x),x)-k(g_n(x),x)|$$

$$= \lim_{n\to\infty} \max\{|k(f_m(x),x)-k(g_m(x),x)|;\ n \le m \le m(n)\}.$$

Let E_n denote the set of all x such that

$$\max\{|k(f_m(x),x)-k(g_m(x),x)|;\ n \le m \le m(n)\} > \epsilon/2$$

By definition, $E \subset \cup_{p=1}^{\infty} \cap_{n=p}^{\infty} E_n$. Therefore there is an integer p such that

$$|\bigcap_{n=p}^{\infty} E_n| > 0.$$

It is easy to see that we can find f^n and g^n both bounded by M such that for almost all x

$$|k(f^n(x),x)-k(g^n(x),x)| = \max\{|k(f_m(x),x)-k(g_m(x),x)|;\ n \le m \le m(n)\}$$

and

$$|f^n(x)-g^n(x)| \le \max\{|f_m(x)-g_m(x)|;\ n \le m \le m(n)\}.$$

Then $\|f^n-g^n\|_\infty \to 0$ as $n \to \infty$ and for $x \in \cap_{n=p}^{\infty} E_n$ and $n = p,p+1,\ldots$ we have

$$|k(f^n(x),x) - k(g^n(x),x)| > \epsilon/2.$$

Applying Lemma 14.3 again, we can find q(n) such that

$$u(x) = \limsup_{n\to\infty} f^n(x) = \lim_{n\to\infty} \max\{f^q(x);\ n \le q \le q(n)\}.$$

By Lemma 14.5, given

$$u_n(x) = \max\{f^q(x);\ n \le q \le q(n)\}$$

we can find v_n such that for almost all x

$$|u_n(x)-v_n(x)| \le \max\{|f^q(x)-g^q(x)|;\ n \le q \le q(n)\}$$

$$\min\{|k(f^q(x),x)-k(g^q(x),x)|;\ n \le q \le q(n)\} \le |k(u_n(x),x)-k(v_n(x),x)|,$$

Finally, we obtain $\|u_n-v_n\|_\infty \to 0$ as $n \to \infty$ and

$$\lim_{n\to\infty} u_n(x) = u(x) = \lim_{n\to\infty} v_n(x)$$

almost everywhere. By Lemma 14.4 and the fact that T is boundedly continuous we obtain

$$\lim_{n\to\infty} \int_a^b k(u_n(x),x)dx = \lim_{n\to\infty} T(u_n) = T(u).$$

Similarly, we have

$$\lim_{n\to\infty} \int_a^b k(v_n(x),x)dx = \lim_{n\to\infty} T(v_n) = T(u).$$

Following the same argument as in the proof of Lemma 14.6, we put

$$E_n^+ = \{x; k(u_n(x),x)-k(v_n(x),x) \ge 0\},\quad E_n^- = [a,b]-E_n^+.$$

As $n \to \infty$ we obtain

$$u_n\chi_{E_n^+} v_n\chi_{E_n^-} \to u \quad \text{almost everywhere},$$

$$v_n\chi_{E_n^+} u_n\chi_{E_n^-} \to u \quad \text{almost everywhere},$$

and consequently

$$\int_a^b |k(u_n(x),x)-k(v_n(x),x)|dx$$

$$= T(u_n\chi_{E_n^+} + v_n\chi_{E_n^-})-T(v_n\chi_{E_n^+} + u_n\chi_{E_n^-})\to 0.$$

It follows that $|k(u_n(x),x)-k(v_n(x),x)|$ converges in measure to 0 as $n \to \infty$. This contradicts the fact that for $x \in \cap_{n=p}^{\infty} E_n$ and $n = p,p+1,\ldots$

$$|k(u_n(x),x) - k(v_n(x),x)| \geq |k(f^n(x),x) - k(g^n(x),x)| > \epsilon/2.$$

Hence for every $\epsilon > 0$ we have $|E| = 0$. That is, the required result holds.

Thereom 14.8. Let T be a boundedly continuous and orthogonally additive functional on L_∞, $|f_i(x)| \leq M$ and $|g_i(x)| \leq M$ for $i = 1,2,\ldots$ and all $x \in [a,b]$. If f_i and g_i are simple functions satisfying $\|f_i-g_i\|_\infty \to 0$ as $i \to \infty$ then

$$T(f)-T(g) \to 0 \text{ as } i \to \infty.$$

Proof. In view of Lemma 14.4, we have

$$T(f_i) = \int_a^b k(f_i(x),x)dx$$

for $i = 1,2,\ldots$, and similarly with f_i replaced by g_i, where k is the kernel function as defined in Lemma 14.4. Furthermore, both $\{k(f_i(x),x)\}$ and $\{k(g_i(x),x)\}$ are dominated by an absolutely Henstock integrable function on [a,b], by Lemma 14.6. It follows from the assumption and Lemma 14.7 that the sequence $\{k(f_i(x),x) - k(g_i(x),x)\}$ is dominated and pointwise convergent to 0 and therefore

$$T(f_i)-T(g_i) \to 0 \quad \text{as } i \to \infty.$$

We remark that the above theorem was proved for simple functions f_i and g_i. Indeed, it also holds true for bounded measurable functions. However, what is required later is for step functions only. The major difficulty in Theorem 14.8 is that when $\|f_i-g_i\|_\infty \to 0$ as $i \to \infty$

we do not necessarily have either f_i or g_i boundedly convergent to a limit function. Therefore we have to go via the kernel function in order to prove the result.

Theorem 14.9. Let T be a functional defined on L_∞. Then T is boundedly continuous and orthogonally additive if and only if there is a nonlinear measure ϕ satisfying (N1) to (N5) such that f is ϕ-integrable on [a,b] whenever $f \in L_\infty$ and

$$T(f) = \int_a^b \phi(f(\cdot),\cdot) \qquad \text{for } f \in L_\infty.$$

Proof. We have proved the sufficiency in Theorem 14.2. Now we prove the necessity. Let $\phi(s,I) = T(s\chi_I)$ where χ_I is the characteristic function of I. Since T is orthogonally additive, ϕ satisfies (N1) and (N3). Since T is boundedly continuous, for any sequence of step functions $\{\varphi_n\}$ boundedly convergent to 0 we have

$$T(\varphi_n) \to 0 \quad \text{as } n \to \infty.$$

Hence (N5) is satisfied. Next, (N4) follows from Theorem 14.8 by taking f_i and g_i to be step functions. Note that (N2) is a special case of (N4). That is, ϕ satisfies conditions (N1) to (N5).

It is obvious that for any step function φ we have

$$T(\varphi) = \int_a^b \phi(\varphi(\cdot),\cdot).$$

Given $f \in L_\infty$, there is a sequence of step functions $\{\varphi_n\}$ boundedly convergent to f. As in the proof of Theorem 14.2, the ϕ-primitives $\Phi_{\phi,n}$ of φ_n are uniformly absolutely continuous. Therefore

$$\begin{aligned} T(f) &= \lim_{n\to\infty} T(\varphi_n) \\ &= \lim_{n\to\infty} \int_a^b \phi(\varphi_n(\cdot),\cdot) \\ &= \int_a^b \phi(f(\cdot),\cdot). \end{aligned}$$

The proof is complete.

We remark that the above functional may also be expressed in terms of a kernel function k as given in Lemma 14.4. It follows from

Lemma 14.7 that $k(\cdot,x)$ is continuous on the real line for almost all $x \in [a,b]$. Hence Theorem 14.9 can be re-stated as follows.

Theorem 14.10. Let T be a functional defined on L_∞. Then T is boundedly continuous and orthogonally additive if and only if $k(f(\cdot),\cdot)$ is absolutely Henstock integrable on $[a,b]$ whenever $f \in L_\infty$ and

$$T(f) = \int_a^b k(f(x),x)dx \quad \text{for } f \in L_\infty$$

where $k(0,x) = 0$, $k(\cdot,x)$ is continuous for almost all x, and $k(s,\cdot)$ is measurable for all s.

If T is boundedly continuous and linear, then $k(s,x) = sk(x)$ in Theorem 14.10 and

$$T(f) = \int_a^b f(x)k(x)dx$$

for all $f \in L_\infty$ and some absolutely Henstock integrable k.

15. ORTHOGONALLY ADDITIVE FUNCTIONALS ON THE DENJOY SPACE

As usual, we consider first the absolute case and then the nonabsolute case. Let L denote the space of all Lebesgue or absolutely Henstock integrable functions on $[a,b]$. The norm in L is given by

$$\|f\|_1 = \int_a^b |f(x)|dx.$$

A continuous functional T on L here means "norm continuous". It is equivalent to the following : if $f_n(x) \to f(x)$ almost everywhere as $n \to \infty$ and $|f_n(x)| \le g(x)$ for almost all x in $[a,b]$ with $g \in L$, then

$$T(f_n) \to T(f) \text{ as } n \to \infty.$$

By definition, we write

$$\int_E \phi(f(\cdot),\cdot) = \int_a^b \phi((f\chi_E)(\cdot),\cdot),$$

if the latter exists, where χ_E denotes the characteristic function of E. We shall use this notation in the following lemma and later in Lemma 15.6.

Lemma 15.1. Let ϕ satisfies (N1) to (N5) such that if f is absolutely Henstock integrable on $[a,b]$ then f is ϕ-integrable on $[a,b]$.

(a) If F_ϕ is the ϕ-primitive of $f \in L$, then F_ϕ is absolutely continuous on $[a,b]$.

(b) If $F_{\phi,n}$ are the ϕ-primitives of $f_n \in L$ and the primitives F_n of f_n are uniformly absolutely continuous, then so are $F_{\phi,n}$.

Proof. Let F_ϕ be the ϕ-primitive of $f \in L$, Define $f_n \in L_\infty$ such that $|f_n(x)| \le |f(x)|$ for almost all $x \in [a,b]$ and all n, and that $f_n(x) \to f(x)$ almost everywhere as $n \to \infty$. Obviously, the primitives F_n of f_n are uniformly absolutely continuous on $[a,b]$. As shown in the proof of Theorem 14.2, each ϕ-primitive $F_{\phi,n}$ of f_n is absolutely continuous. We want to show that $F_{\phi,n}$ are uniformly absolutely continuous. If so, then F_ϕ being the limit of $F_{\phi,n}$ is also absolutely continuous on $[a,b]$.

Suppose false, i.e., there exists $\epsilon > 0$ such that for every i we can find a measurable set E_i and a function $f_{n(i)}$ satisfying

$$|E_i| < 2^{-i} \quad \text{and} \quad \left| \int_{E_i} \phi(f_{n(i)}(\cdot),\cdot) \right| \ge 2\epsilon.$$

For convenvience, we denote the above integral by $F_{\phi,n(i)}(E_i)$, which is well-defined if $f_{n(i)}$ and $|f_{n(i)}|$ are ϕ-integrable on $[a,b]$. Since each $F_{\phi,n}$ is absolutely continuous, we may choose E_{i+1} after E_i so that

$$|F_{\phi,n(i)}(E_{i+1})| < \epsilon.$$

Then it follows that

$$|F_{\phi,n(i)}(E_i - E_{i+1})| \ge |F_{\phi,n(i)}(E_i)| - |F_{\phi,n(i)}(E_i \cap E_{i+1})| > \epsilon.$$

Note that $E_i \cap E_{i+1}$ may be empty. Now put $g(x) = f_{n(i)}(x)$ when $x \in E_i - E_{i+1}$ for all i and 0 otherwise. Then $g \in L$ and

$$\int_a^b \phi(g(\cdot),\cdot) = \sum_{i=1}^{\infty} F_{\phi,n(i)}(E_i - E_{i+1})$$

which diverges. It contradicts the assumption that g is ϕ-integrable on $[a,b]$. Therefore $F_{\phi,n}$ are uniformly absolutely continuous. We have proved (a).

Using (a), repeat the above argument and we obtain (b).

We remark that the above lemma remains valid if $f_n \in L_\infty$ for all n in (b) and if the ϕ-integrability of f is replaced by the following condition: T is a continuous and orthogonally additive functional on L with

$$T(f) = \int_a^b \phi(f(\cdot),\cdot) \qquad \text{for } f \in L_\infty.$$

Theorem 15.2. Let T be a functional defined on L. Then T is continuous and orthogonally additive if and only if there is a nonlinear measure ϕ satisfying (N1) to (N5) such that f is ϕ-integrable on [a,b] whenever $f \in L$ and

$$T(f) = \int_a^b \phi(f(\cdot),\cdot) \qquad \text{for } f \in L.$$

Proof. Suppose T is continuous and orthogonally additive on L. When restricted to L_∞ the functional T is boundedly continuous on L_∞. Then by Theorem 14.9 there is a nonlinear measure ϕ satisfying (N1) to (N5) such that

$$T(f) = \int_a^b \phi(f(\cdot),\cdot) \quad \text{for } f \in L_\infty.$$

Let $f \in L$. Then there is a sequence of bounded functions f_n converging almost everywhere to f and dominated by f. Hence by Lemma 15.1(b) and the remark thereafter we have

$$\begin{aligned} T(f) &= \lim_{n\to\infty} T(f_n) \\ &= \lim_{n\to\infty} \int_a^b \phi(f_n(\cdot),\cdot) \\ &= \int_a^b \phi(f(\cdot),\cdot). \end{aligned}$$

Conversely, let ϕ satisfy (N1) to (N5). Again, applying Lemma 15.1(b) we see that T is continuous. The orthogonal additivity follows from (N1). The proof is complete.

Following Theorem 14.10, we can re-state the above theorem as follows.

Theorem 15.3. Let T be a functional defined on L. Then T is continuous and orthogonally additive if and only if $k(f(\cdot),\cdot)$ is absolutely Henstock integrable on [a,b] whenever $f \in L$ and

$$T(f) = \int_a^b k(f(x),x)dx \qquad \text{for } f \in L$$

where $k(0,x) = 0$, $k(\cdot,x)$ is continuous for almost all x, and $k(s,\cdot)$ is measurable for all s.

If T is linear, then $k(s,x) = sk(x)$ in Theorem 15.3 and applying Theorem 12.8 we have

Theorem 15.4. A functional T is continuous and linear on L if and only if

$$T(f) = \int_a^b f(x)k(x)dx$$

for all $f \in L$ and some $k \in L_\infty$.

To proceed further, we need the following lemmas.

Lemma 15.5. Let f be Henstock integrable on $[a,b]$ with the primitive F and X a closed subset in $[a,b]$ with

$$(a,b)-X = \bigcup_{k=1}^{\infty} (a_k,b_k).$$

Then F is $AC^*(X)$ if and only if the following conditions are satisfied:

(i) for every $\epsilon > 0$ there exists an $\eta > 0$ such that whenever $E \subset X$ and $|E| < \eta$ we have

$$\left|\int_E f(x)dx\right| < \epsilon;$$

(ii) $\sum_{k=1}^{\infty} \omega(F;[a_k,b_k])$ converges where ω denotes the oscillation.

Proof. Suppose F is $AC^*(X)$. Then it follows from Lemma 6.18 that f is Henstock integrable on X and indeed on any measurable subset E of X. Let $f_X(x) = f(x)$ when $x \in X$ and 0 otherwise. Then the primitive of f_X is absolutely continuous. It is easy to verify that for every $\epsilon > 0$ there is an $\eta > 0$ such that whenever $|E| < \eta$

$$\left|\int_E f_X(t)dt\right| < \epsilon.$$

Hence (i) holds. Condition (ii) is an immediate consequence of the definition of $AC^*(X)$.

Conversely, suppose (i) and (ii) are satisfied. In view of Lemma 6.4, it is sufficient to show that F is $AC(X)$. For every $\epsilon > 0$ there is an integer N such that

$$\sum_{k=N+1}^{\infty} \omega(F;\ [a_k,b_k]) < \epsilon.$$

Choose $\eta > 0$ as in (i) and also $\eta < |b_k - a_k|$ for $k = 1,2,\ldots,N$. Then whenever $\sum_i |I_i| < \eta$ with $I_1, I_2, \ldots$ nonoverlapping and the endpoints of each I_i belonging to X we have

$$\sum_i |F(I_i)| \le \sum_i \left| \int_{I_i \cap X} f \right| + \sum_i \left| \int_{I_i - X} f \right|$$
$$< 3\epsilon.$$

The proof is complete.

Lemma 15.6. Let f_n be ϕ-integrable on [a,b] with ϕ-primitive $F_{\phi,n}$ for $n = 1,2,\ldots$ and X a closed subset in [a,b] with

$$(a,b) - X = \bigcup_{k=1}^{\infty} (a_k, b_k).$$

Then $F_{\phi,n}$ are $AC^*(X)$ uniformly in n if and only if the following conditions are satisfied:

(i) for every $\epsilon > 0$ there exists an $\eta > 0$ such that whenever $E \subset X$ and $|E| < \eta$

$$\left| \int_E \phi(f_n(\cdot),\cdot) \right| < \epsilon \quad \text{for all } n;$$

(ii) $\sum_{k=1}^{\infty} \omega(F_{\phi,n}; [a_k,b_k])$ converges uniformly in n.

Proof. Suppose $F_{\phi,n}$ are $AC^*(X)$ uniformly in n. Then following the same argument as in Lemma 6.18 we can show that each f_n is ϕ-integrable on X and furthermore ϕ-integrable on any measurable subset E of X.

Consider the inequality

$$\left| \int_E \phi(f_n(\cdot),\cdot) \right| \le \left| \int_E \phi(f_n(\cdot),\cdot) - \sum \phi(f_n(\xi),I) \right|$$
$$+ \left| \sum \phi(f_n(\xi),I) - \sum F_{\phi,n}(I) \right| + \left| \sum F_{\phi,n}(I) \right|$$

where $\sum$ sums over a partial division $\{[u,v];\xi\}$ with $\xi \in E$. Given $\epsilon > 0$, we may choose the partial division suitably so that each of the first two terms on the right side of the above inequality is less than ϵ. Hence we have proved (i). Again, condition (ii) is an immediate consequence of the definition of $AC^*(X)$.

The sufficiency follows as in the proof of Lemma 15.5.

Lemma 15.7. Let ϕ satisfy (N1) to (N5) such that if a function f is Henstock integrable on [a,b] then it is ϕ-integrable on [a,b], and let each f_n be Henstock and ϕ-integrable on [a,b] with primitives F_n and $F_{\phi,n}$ respectively for n = 1,2,... . If F_n are $AC^*(X)$ uniformly in n, then so are $F_{\phi,n}$.

Proof. First, for each fixed n we shall prove that if F_n is $AC^*(X)$, then so is $F_{\phi,n}$. For convenience, write F and F_ϕ for F_n and $F_{\phi,n}$. Suppose F is $AC^*(X)$ and is the primitive of f. Let $f_X(x) = f(x)$ when $x \in X$ and 0 otherwise. Then by Theorem 6.18 the function f_X is absolutely Henstock integrable on [a,b]. Applying Lemma 15.1(a) with f replaced by f_X, we obtain that condition (i) of Lemma 15.6 is satisfied with f_n replaced by f. Put

$$(a,b)-X = \bigcup_{k=1}^{\infty} (a_k,b_k).$$

It remains to show that

$$\sum_{k=1}^{\infty} \omega(F_\phi;[a_k,b_k]) < +\infty.$$

Since F_ϕ is continuous, for some $[c_k,d_k] \subset [a_k,b_k]$ we have

$$\omega(F_\phi;\ [a_k,b_k]) = |F_\phi(c_k,d_k)|$$

Put $g(x) = f(x)$ if $x \in [c_k,d_k]$ for some k and $F_\phi(c_k,d_k) \geq 0$, and 0 otherwise. It follows that g is ϕ-integrable on [a,b] and furthermore

$$\sum_{k\in\sigma} F_\phi(c_k,d_k) = \int_a^b \phi(g(\cdot),\cdot)$$

where σ denotes the set of all k for which $F_\phi(c_k,d_k) \geq 0$. Similarly, the series of all $F_\phi(c_k,d_k)$, for which $F_\phi(c_k,d_k) \leq 0$ is also convergent. Therefore F_ϕ is $AC^*(X)$.

Next, we shall show that $F_{\phi,n}$ are $AC^*(X)$ uniformly in n. Condition (i) in Lemma 15.6 follows from Lemma 15.1(b). It remains to prove Lemma 15.6 (ii). Suppose false, there exists $\epsilon > 0$ such that we can find m(i) and n(i) for i = 1,2,... satisfying

$$\sum_{k=m(i)}^{\infty} \omega(F_{\phi,n(i)};[a_k,b_k]) > \epsilon$$

where (a_k,b_k) are the contiguous intervals of X. Then there is p(i) such that

$$\sum_{k=m(i)}^{p(i)} \omega(F_{\phi,n(i)};[a_k,b_k]) > \epsilon.$$

Let E_i be the union of $[a_k,b_k]$ for $k = m(i),\ldots,p(i)$. We may assume that E_i are pairwise disjoint. Define $g(x) = f_{n(i)}(x)$ when $x \in E_i$, $i = 1,2,\ldots$, and 0 elsewhere. Then g is Henstock integrable and therefore ϕ-integrable on [a,b]. In view of the first part of the proof, writing G_ϕ as the ϕ-primitive of g we have

$$\sum_{k=1}^{\infty} \omega(G_\phi;[a_k,b_k]) = \sum_{i=1}^{\infty} \sum_{k=m(i)}^{p(i)} \omega(F_{\phi,n(i)};[a_k,b_k]) < +\infty$$

which leads to a contradiction. Hence Lemma 15.6 (ii) holds and $F_{\phi,n}$ are $AC^*(X)$ uniformly in n.

We remark that the above lemma remains valid if $f_n \in L$ for all n and the ϕ-integrability of f is replaced by the following condition: T is control-continuous and orthogonally additive on the Denjoy space E with

$$T(f) = \int_a^b \phi(f(\cdot),\cdot) \quad \text{for } f \in L.$$

The same remark will also apply to Lemmas 15.8 and 15.9. This remark will be used in the proof of Theorem 15.10.

We recall that the $UACG^*$ invariance and equicontinuity invariance of ϕ were defined in Definition 13.10. As an immediate consequence of Lemma 15.7, we have

Lemma 15.8. If ϕ satisfies conditions (N1) to (N5) such that if a function f is Henstock integrable on [a,b] then it is ϕ-integrable on [a,b], then ϕ is $UACG^*$ invariant.

Lemma 15.9. If ϕ satisfies conditions (N1) to (N5) such that if f is Henstock integrable on [a,b] then it is ϕ-integrable on [a,b], then ϕ is equicontinuity invariant.

Proof. Let f_n, $n = 1,2,\ldots$, be Henstock and ϕ-integrable on [a,b] with primitives F_n and $F_{\phi,n}$ respectively. Suppose F_n are equicontinuous and yet $F_{\phi,n}$ are not. Then for every $x \in [a,b]$ and every j there exists a $\delta_j > 0$ such that whenever $|y-x| < \delta_j$

$$|F_n(y) - F_n(x)| < 2^{-j} \quad \text{for all } n.$$

On the other hand, there exist $x_o \in [a,b]$ and $\epsilon > 0$ such that for every $\delta_j > 0$ there exists y_j with $|x_o - y_j| < \delta_j$ and there exists $n(j)$ satisfying

$$|F_{\phi,n(j)}(x_o) - F_{\phi,n(j)}(y_i)| > 2\epsilon.$$

Since $F_{\phi,n(j)}$ is continuous at x_o, there is z_j which lies between x_o and y_j such that

$$|F_{\phi,n(j)}(x_o) - F_{\phi,n(j)}(z_j)| < \epsilon$$

and therefore

$$|F_{\phi,n(j)}(y_j) - F_{\phi,n(j)}(z_j)| > \epsilon.$$

Let I_j denote the closed interval with endpoints y_i and z_j. Note that y_i and z_j lie on the same side of x_o. Further, we may choose I_j so that they are pairwise disjoint. In view of the assumption, we have

$$\sum_j \omega(F_{n(j)}; I_j) < +\infty.$$

Take $f(x) = f_{n(j)}(x)$ when $x \in I_j$ and 0 otherwise. Then f has the primitive F such that $F(u,v) = F_{n(j)}(u,v)$ when $[u,v] \subset I_j$. Thus by Lemma 15.7 we obtain

$$\sum_j \omega(F_\phi; I_j) = \sum_j \omega(F_{\phi,n(j)}; I_j) < +\infty.$$

It leads to a contradiction. Hence $F_{\phi,n}$ are equicontinuous.

Theorem 15.10. Let T be a functional defined on the Denjoy space E. Then T is control-continuous and orthogonally additive if and only if there is a nonlinear measure ϕ satisfying (N1) to (N5) such that f is ϕ-integrable on [a,b] whenever $f \in E$ and

$$T(f) = \int_a^b \phi(f(\cdot),\cdot) \qquad \text{for } f \in E.$$

Proof. It follows the same argument as in the proof of Theorem 15.2. Suppose T is control-continuous and orthogonally additive on E. Then when restricted to L it is norm continuous on L. It follows from Theorem 15.2 that $\phi(s,I) = T(s\chi_I)$ satisfies (N1) to (N5) and

$$T(f) = \int_a^b \phi(f(\cdot),\cdot) \qquad \text{for } f \in L.$$

Let $f \in E$. Then there is a sequence of functions $f_n \in L$ control-convergent to f. Applying Lemmas 15.8 and 15.9 and the controlled convergence theorem for the ϕ-integral (Theorem 13.6) and keeping in mind the remark after Lemma 15.7 we obtain

$$T(f) = \int_a^b \phi(f(\cdot),\cdot).$$

Conversely, let f_n be control-convergent to f. Repeat the above argument and we have

$$T(f_n) \to T(f) \quad \text{as } n \to \infty.$$

The orthogonal additivity follows from (N1) as usual. Hence the proof is complete.

Again, following Theorem 14.10, we can re-state the above theorem as follows.

Theorem 15.11. Let T be a functional defined on the Denjoy space E. Then T is control-continuous and orthogonally additive if and only if $k(f(\cdot),\cdot)$ is Henstock integrable on [a,b] whenever $f \in E$ and

$$T(f) = \int_a^b k(f(x),x)dx \qquad \text{for } f \in E$$

where $k(0,x) = 0$, $k(\cdot,x)$ is continuous for almost all x, and $k(s,\cdot)$ is measurable for all s.

If T is linear, then $k(s,x) = sk(x)$ in Theorem 15.11 and applying Theorem 12.9 we have

Theorem 15.12. A functional T is control-continuous and linear on the Denjoy space E if and only if

$$T(f) = \int_a^b f(x)k(x)dx$$

for all $f \in E$ and some k of bounded variation on [a,b].

Comparing with Theorem 12.7, we see that the norm continuity and control-continuity of a linear functional on the Denjoy space are equivalent.

CHAPTER 4 THE RIEMANN-TYPE INTEGRALS

16. THE RL INTEGRAL AND THE MCSHANE INTEGRAL

In this chapter, we shall introduce other kinds of Riemann-type integrals. The Henstock integral uses a variable $\delta(\xi)$. However, for computing the value of an integral we would still prefer a constant δ. This provides the motivation for introducing the following RL integral. Here RL stands for Riemann-Lebesgue. For convenience, given a division $D = \{[u,v];\xi\}$ we write

$$\sum_{P} f(\xi)(v-u)$$

for the sum over D in which the property P holds.

Definition 16.1. A non-negative function f is said to be RL integrable on [a,b] if there exists a number A such that for every $\epsilon > 0$ and $\eta > 0$ there exist an open set G and a constant $\delta > 0$ such that $|G| < \eta$ and that for every division $D = \{[u,v];\xi\}$ with $0 < v - u < \delta$ and $\xi \in [u,v] - G$ we have

$$\Big| \sum_{\xi \notin G} f(\xi)(v - u) - A \Big| < \epsilon.$$

It is understood that the term $f(\xi)(v - u)$ is not included when $[u,v] - G$ is empty. In general, a function f is said to be RL integrable on [a,b] if both $f^+ = \max(f,0)$ and $f^- = \max(-f,0)$ are RL integrable on [a,b], and we write

$$\int_a^b f = \int_a^b f^+ - \int_a^b f^-.$$

The uniqueness of the RL integral follows from Theorem 16.3 below.

Lemma 16.2. Let f be bounded on [a,b]. Then f is RL integrable on [a,b] if and only if f is absolutely Henstock integrable on [a,b].

Proof. We may assume that f is non-negative. Suppose f is RL integrable on [a,b]. Then given $\epsilon > 0$ and $\eta > 0$ there exist an open set G and a constant $\delta > 0$ such that $|G| < \eta$ and the rest of the condition holds. Define $\delta(\xi) = \delta/2$ when $\xi \notin G$ and $(\xi - \delta(\xi),$

$\xi + \delta(\xi)) \subset G$ when $\xi \in G$. Then we see that f is Henstock integrable on [a,b].

Conversely, suppose f is bounded by N, non-negative and Henstock integrable on [a,b]. Then for every $\epsilon > 0$ there is a $\delta(\xi) > 0$ such that for any δ-fine division $D = \{[u,v];\xi\}$ we have

$$\sum |f(\xi)(v - u) - F(u,v)| < \epsilon.$$

Since $\delta(\xi)$ is measurable by Theorem 10.3, given $\eta > 0$ we may choose $\delta > 0$ and an open set G such that $G \supset \{\xi; 0 < \delta(\xi) < \delta\}$ and $|G| < \eta$. Hence for any division $D = \{[u,v];\xi\}$ with $0 < v - u < \delta$ and $\xi \in [u,v] - G$ we have

$$\Big|\sum_{\xi \notin G} f(\xi)(v-u) - F(a,b)\Big| \le \Big|\sum_{\xi \notin G} \{f(\xi)(v-u) - F(u,v)\}\Big| + \Big|\sum_{\xi \in G} F(u,v)\Big| < \epsilon + N\eta.$$

That is, f is RL integrable on [a,b].

We remark that going through the previous proof again we see that a bounded function f is RL integrable on [a,b] if and only if the condition in Definition 16.1 holds. In other words, when f is bounded there is no need to split the function into two non-negative components f^+ and f^-.

Theorem 16.3. A function is RL integrable on [a,b] if and only if it is absolutely Henstock integrable there.

Proof. Again, we may assume that f is non-negative. Suppose f is RL integrable to A on [a,b]. Let f^N denote the truncated function of f, i.e., $f^N(x) = f(x)$ when $f(x) \le N$ and $f^N(x) = N$ when $f(x) > N$. We shall show that f^N is Henstock integrable on [a,b].

Suppose the condition in Definition 16.1 holds. Define $\delta(\xi) = \delta/2$ when $\xi \notin G$ and $(\xi - \delta(\xi), \xi + \delta(\xi)) \subset G$ when $\xi \in G$. Take any two δ-fine divisions $D_1 = \{[u',v'];\xi'\}$ and $D_2 = \{[u'',v''];\xi''\}$. We want to show that

$$\Big|\sum_1 f^N(\xi')(v'-u') - \sum_2 f^N(\xi'')(v''-u'')\Big| < \epsilon$$

where $\sum_1$ sums over D_1 and $\sum_2$ over D_2. If so, by Lemma 3.4 the function f^N is Henstock integrable on [a,b].

In particular, when D_2 is finer than D_1 we write

$$\sum_2 \omega(f^N;[u',v'] - G)(v''-u'')$$

to mean that $\sum_2$ sums over D_2 whereas the oscillation ω is taken over $[u',v'] - G$ with $[u',v']$ in D_1 and $[u',v'] \supset [u'',v'']$. It is understood that $\omega = 0$ when $[u',v'] \subset G$. This is meaningful because D_2 is finer than D_1. Then we have

$$\begin{aligned}|\textstyle\sum_1 f^N(\xi')(v'-u') - \sum_2 f^N(\xi'')(v''-u'')| &\le \textstyle\sum_2 \omega(f^N;[u',v'] - G)(v''-u'') + 2N\eta \\ &\le \textstyle\sum_1 \omega(f^N;[u',v']-G)(v'-u') + 2N\eta \\ &\le \textstyle\sum_1 \omega(f;[u',v']-G)(v'-u') + 2N\eta \\ &< 2\epsilon + 2N\eta.\end{aligned}$$

In general, take another δ-fine division $D_3 = \{[u,v];\xi\}$ finer than both D_1 and D_2. It follows that

$$\begin{aligned}|\textstyle\sum_1 f^N(\xi')(v'-u') - \sum_2 f^N(\xi'')(v''-u'')| &\le |\textstyle\sum_1 - \sum_3| + |\sum_3 - \sum_2| \\ &< 4\epsilon + 4N\eta.\end{aligned}$$

Hence we have proved that f^N is Henstock integrable on $[a,b]$.

Since $f^N(x)$ converges monotonely increasing to $f(x)$ as $N \to \infty$ and

$$\int_a^b f^N(x)dx \le A \quad \text{for all } N,$$

then by the monotone convergence theorem (Theorem 4.1) for the Henstock integral, f is Henstock integrable on $[a,b]$.

Conversely, suppose f is non-negative and Henstock integrable on $[a,b]$. Then by Lemma 5.4 the truncated function f^N is also Henstock integrable on $[a,b]$. Given $\epsilon > 0$ and $\eta > 0$, there are an integer N and an open set $E_N \supset \{x;\ f(x) > N\}$ such that $N|E_N| < \eta/2$ and

$$\left|\int_a^b f(x)dx - \int_a^b f^N(x)dx\right| < \epsilon.$$

It follows from Lemma 16.2 that f^N is RL integrable on $[a,b]$. Let A_N denote the integral of f^N on $[a,b]$. Then there are an open set G_N and a constant $\delta > 0$ such that $N|G_N| < \eta/2$ and that for any division $D = \{[u,v];\xi\}$ with $0 < v - u < \delta$ and $\xi \in [u,v] - G_N$ we have

$$\left|\sum_{\xi \notin G_N} f^N(\xi)(v - u) - A_N\right| < \epsilon.$$

Write $G = E_N \cup G_N$ and note that $f^N(x) = f(x)$ when $x \notin G$. Then $N|G| < \eta$, in particular, $|G| < \eta$ and

$$\left| \sum_{\xi \notin G} f(\xi)(v-u) - A \right| \le |A - A_N| + \left|A_N - \sum_{\xi \notin G} f^N(\xi)(v-u)\right|$$
$$< \epsilon + \epsilon + N\eta.$$

The above is so because if $[u,v] - G$ is nonempty then $f^N(\xi)(v-u)$ is one of the terms in the sum over $\xi \notin G_N$, and if $[u,v] - G$ is empty, i.e. $[u,v] \subset G$, then $[u,v] - G_N$ may or may not be empty, the collection of all terms $f^N(\xi)(v-u)$, when $[u,v] - G_N$ is nonempty, is less than $N|G|$ or $N\eta$. Consequently, f is RL integrable on $[a,b]$.

In fact, we have also proved the following

Corollary 16.4. Let f be non-negative and RL integrable on $[a,b]$. Then we may choose G in the definition of the RL integral so that $G \supset \{x;\ f(x) > N\}$ for some positive integer N and $N|G| < \eta$.

Example 16.5. Let $f(x) = 1/\sqrt{x}$ for $0 < x \le 1$ and $f(0) = 0$. Example 2.7 shows that f is Henstock integrable on $[0,1]$. As an easy consequence of the definition of the RL integral, we obtain

$$\lim_{n\to\infty} \frac{1}{n} \sum_{i=1}^{n} f\left(\frac{i}{n}\right) = \int_0^1 f(x)dx.$$

This provides a means to compute the value of the integral.

Example 16.6. Let

$$f(x) = 2n^2(n+1) \quad \text{when} \quad x \in \left(\frac{1}{2}\left(\frac{1}{n+1} + \frac{1}{n}\right), \frac{1}{n}\right),$$
$$= -2n^2(n+1) \quad \text{when} \quad x \in \left(\frac{1}{n+1}, \frac{1}{2}\left(\frac{1}{n+1} + \frac{1}{n}\right)\right),\ n = 1,2,\ldots,$$

and $f(x) = 0$ elsewhere in $[0,1]$. Then with $A = 0$ and for $\epsilon > 0$ and $\eta > 0$ we can choose $G = [0,1/n)$ with $1/n < \eta$ and $\delta > 0$ such that for any division $D = \{[u,v];\xi\}$ with $0 < v - u < \delta$ we have

$$\left| \sum_{\xi \notin G} f(\xi)(v-u) - A \right| < \epsilon.$$

In fact, given any real number A, it is easy to find an open set G (depending on A) such that the above inequality holds. This shows that f being non-negative in the definition of the RL integral is essential. The above function f is not even Henstock integrable on $[0,1]$.

Here we introduce another absolute integral, namely the Mcshane integral. We shall show that the RL, the absolutely Henstock, and the McShane integrals are all equivalent.

Definition 16.7. A function f is said to be McShane integrable on [a,b] if there is a number A such that for every $\epsilon > 0$ there exists a $\delta(\xi) > 0$ such that for every division $D = \{[u,v];\xi\}$ of [a,b] satisfying $[u,v] \subset (\xi - \delta(\xi), \xi + \delta(\xi))$, we have

$$\left|\sum f(\xi)(v-u) - A\right| < \epsilon.$$

Note that in the above definition we do not require $\xi \in [u,v]$. In other words, ξ may lie outside [u,v], For such a division we call it a δ-fine McShane division.

Example 16.8. Every continuous function is McShane integrable. It is instructional to go through the proof of Example 2.6 again. We find that the proof remains valid with δ-fine divisions replaced by δ-fine McShane divisions. Note that it does not work for Example 2.4.

Theorem 16.9. If f is McShane integrable on [a,b] then so is $|f|$.

Proof. Given $\epsilon > 0$ there is a $\delta(\xi) > 0$ such that for any δ-fine McShane division $D = \{[u,v];\xi\}$ we have

$$\left|\sum f(\xi)(v-u) - A\right| < \epsilon.$$

Let $D' = \{[u',v'];\xi'\}$ and $D'' = \{[u'',v''];\ \xi''\}$ be any two δ-fine McShane divisions with $\sum'$ over D' and $\sum''$ over D''. Denote by [u,v] the intersection $[u',v'] \cap [u'',v'']$ and by $\sum$ the sum over all [u,v]. Then

$$\left|\sum f(\xi')(v-u) - \sum f(\xi'')(v-u)\right| \le \left|\sum' f(\xi')(v'-u') - A\right| + \left|A - \sum'' f(\xi'')(v''-u'')\right| < 2\epsilon.$$

Here we may retain ξ' or ξ'' as the associated point of [u,v]. We observe that Henstock's lemma (Theorem 3.7) holds true for the McShane integral with the same proof. Therefore

$$\sum |f(\xi') - f(\xi'')|(v-u) < 8\epsilon,$$

and consequently,

$$\begin{aligned} &\left|\sum' |f(\xi')|(v'-u') - \sum'' |f(\xi'')|(v''-u'')\right| \\ &\le \sum \big|\, |f(\xi')| - |f(\xi'')| \,\big|(v-u) \\ &\le \sum |f(\xi') - f(\xi'')|(v-u) < 8\epsilon. \end{aligned}$$

As in Lemma 3.4, this is necessary and sufficient for $|f|$ to be McShane integrable on [a,b]. Hence the proof is complete.

The above proof does not work for ordinary δ-fine divisions because when we change from the sum $\sum'$ over $[u',v']$ to the sum $\sum$ over $[u,v]$ using δ-fine McShane divisions we may retain the associated point ξ' for different $[u,v]$ in $[u',v']$ whereas this is not possible for ordinary δ-fine divisions. As a corollary of Theorem 16.9, we see that every McShane integrable function is absolutely Henstock integrable.

Theorem 16.10. If f is Henstock integrable on [a,b] and non-negative, then f is McShane integrable on [a,b].

Proof. Suppose f is bounded by M. Applying Theorem 5.9 and the proof of Corollary 4.5, we obtain a sequence of continuous functions converging to f almost everywhere. Then it follows from Egoroff's theorem (Lemma 7.2) that for every $\epsilon > 0$ and $\eta > 0$ there exist a continuous function g, also bounded by M, and an open set E with $|E| < \eta$ such that

$$|f(x) - g(x)| < \epsilon \quad \text{for} \quad x \in [a,b] - E,$$
$$|F(a,b) - G(a,b)| < \epsilon,$$

where $F(a,b)$ and $G(a,b)$ denote respectively the integrals of f and g on $[a,b]$. Since g is McShane integrable on [a,b], there exists a $\delta(\xi) > 0$ such that for any δ-fine McShane division $D = \{[u,v],\xi\}$ we have

$$| \sum f(\xi)(v-u) - G(a,b)| < \epsilon.$$

We may assume $(\xi - \delta(\xi),\ \xi + \delta(\xi)) \subset E$ when $\xi \in E$. Then for any δ-fine McShane division $D = \{[u,v];\xi\}$ we have

$$\begin{aligned} | \sum f(\xi)(v-u) - F(a,b)| \le\ & |\sum_1 f(\xi)(v-u) - \sum_1 g(\xi)(v-u)| \\ & +|\sum_1 g(\xi)(v-u) - \sum_1 G(u,v)| + |G(a,b) - F(a,b)| \\ & + |\sum_2 f(\xi)(v-u)| + |\sum_2 G(u,v)| \end{aligned}$$

where $\sum_1$ denotes the partial sum of $\sum$ for which $\xi \in [a,b] - E$ and $\sum_2 = \sum - \sum_1$. Hence using Henstock's lemma for the McShane integral as in the proof of Theorem 16.9, we obtain

$$|\sum f(\xi)(v-u) - F(a,b)| < \epsilon(b-a) + 3\epsilon + 2M\eta.$$

That is, f is McShane integrable on [a,b] provided f is bounded.

In general, put $f_n(x) = f(x)$ when $f(x) \le n$ and 0 otherwise. We observe that the monotone convergence theorem (Theorem 4.1) holds true for the McShane integral with the same proof. Since f_n is McShane integrable on [a,b] for each n, so is the limit function f.

In conclusion, the RL integral or the McShane integral provides an alternative definition to the absolutely Henstock integral, and therefore the Lebesgue integral. The results that hold true for the absolutley Henstock integral in Chapters 1 and 2 also hold for the above two integrals. Another absolute integral will be introduced at the end of Section 17.

17. LOCALLY SMALL RIEMANN SUMS

We shall define locally small Riemann sums or, in short, LSRS and show that it is the necessary and sufficient condition for f to be Henstock integrable on [a,b].

Definition 17.1. A measurable function f defined on [a,b] is said to have LSRS if for every $\epsilon > 0$ there is a $\delta(\xi) > 0$ such that for every $t \in [a,b]$ we have

$$\left|\sum f(\xi)(v-u)\right| < \epsilon$$

whenever $D = \{[u,v];\xi\}$ is a δ-fine division of an interval [r,s] lying in $(t-\delta(t),\ t+\delta(t))$ and $\sum$ sums over D.

Theorem 17.2. If f is Henstock integrable on [a,b] then it has LSRS.

Proof. Let F be the primitive of f. Given $\epsilon > 0$ there is a $\delta(\xi) > 0$ such that for any δ-fine division $D = \{[u,v];\xi\}$ of [a,b] we have

$$\sum\left|f(\xi)(v-u) - F(u,v)\right| < \epsilon.$$

We may assume in addition that

$$|F(u,v)| < \epsilon \quad \text{whenever} \quad [u,v] \subset (\xi-\delta(\xi),\ \xi+\delta(\xi)).$$

Therefore for $t \in [a,b]$ and any δ-fine division $D = \{[u,v];\xi\}$ of $[r,s] \subset (t-\delta(t),\ t+\delta(t))$ we have

$$\begin{aligned}\left|\sum f(\xi)(v-u)\right| &\le \sum\left|f(\xi)(v-u) - F(u,v)\right| + |F(r,s)| \\ &< 2\epsilon.\end{aligned}$$

That is, LSRS is satisfied.

Theorem 17.3. If a measurable function f has LSRS then there is a $\delta(\xi) > 0$ such that the set of all Riemann sums $\sum f(\xi)(v-u)$ over δ-fine divisions $D = \{[u,v];\xi\}$ of [a,b] is bounded.

Proof. Since the family of all $(\xi-\delta(\xi), \xi+\delta(\xi))$ for $\xi \in [a,b]$ is an open cover of the compact interval $[a,b]$, there is a finite subcover, say, $(\xi_i - \delta(\xi_i), \xi_i + \delta(\xi_i))$ for $i = 1,2,\ldots,n$. Hence we may modify $\delta(\xi)$ so that whenever $\xi \in (\xi_i-\delta(\xi_i), \xi_i+\delta(\xi_i))$ for some i we impose the condition on $\delta(\xi)$ that $(\xi-\delta(\xi), \xi+\delta(\xi)) \subset (\xi_i-\delta(\xi_i), \xi_i+\delta(\xi_i))$. Therefore for any δ-fine division $D = \{[u,v];\xi\}$ we have

$$|\sum f(\xi)(v-u)| \le n\epsilon$$

That is, the Riemann sums are bounded.

We remark that Theorem 17.3 holds true because of the compactness of $[a,b]$. Having bounded Riemann sums is not enough to deduce the integrability of a function, as is shown by the following example of Henstock.

Example 17.4. Let $f(x) = 1$ when $0 \le x < 1$, $(-1)^j j(j+1)$ when $2-j^{-1} \le x < 2-(j+1)^{-1}$, $j = 1,2,\ldots$, and $f(2) = 0$. For a suitable $\delta(\xi) > 0$, the corresponding Riemann sums oscillate between 0 and 1 and therefore bounded. However f is not Henstock integrable on $[0,2]$ otherwise

$$\int_0^{2-1/n} f(x)dx = 1 + \sum_{j=1}^{n} (-1)^j$$

would converge as $n \to \infty$ since the Henstock integral is closed under the Cauchy extension (Corollary 7.10).

In view of Theorem 17.3, if f has LSRS then the following $\overline{A}$ and $\underline{A}$ are well-defined:

$$\overline{A} = \inf_{\delta} \sup_{D} \sum f(\xi)(v-u)$$

where the supremum is over all δ-fine divisions $D = \{[u,v];\ \xi\}$ of $[a,b]$ for a given $\delta(\xi)$ and the infimum over all positive functions $\delta(\xi)$; and similarly

$$\underline{A} = \sup_{\delta} \inf_{D} \sum f(\xi)(v-u).$$

To prove $\overline{A} = \underline{A}$, we first define major and minor functions U_δ and V_δ and then apply the category argument as in Theorem 8.12.

Definition 17.5. Suppose f has LSRS. We define

$$U_\delta(x) = \sup_{D} \sum f(\xi)(v-u)$$

where the supremum is over all δ-fine divisions $D = \{[u,v];\xi\}$ of $[a,x]$ for a given $\delta(\xi) > 0$; and similarly

$$V_\delta(x) = \inf_D \sum f(\xi)(v-u).$$

Lemma 17.6. If a measurable function f has LSRS, then U_δ and V_δ are respectively major and minor functions of f on [a,b].

The proof is straight forward. Obviously, in view of Theroem 8.10, both U_δ and V_δ are VBG^* on [a,b]. But they need not be continuous. Therefore we cannot apply the theorem of Marcinkiewicz (Corollary 8.13). However we do have a family of major and minor functions for shrinking $\delta(\xi)$. So a similar technique of proof still applies.

Lemma 17.7. Let f be Henstock integrable on any closed subinterval in (a,b). If f has LSRS on [a,b] then f is Henstock integrable on [a,b].

Proof. Let F(u,v) be the integral of f on $[u,v] \subset (a,b)$. Let U_δ and V_δ be defined as in Definition 17.5. Since f has LSRS, given $\epsilon > 0$ there is a $\delta(\xi) > 0$ such that for $[x,y] \subset (a,a+\delta(a))$ or $[x,y] \subset (b-\delta(b),b)$ we can find a δ-fine division $D = \{[u,v];\xi\}$ of [x,y] with $\sum$ over D satisfying

$$\begin{aligned} |F(x,y)| &\le |F(x,y) - \sum f(\xi)(v-u)| + |\sum f(\xi)(v-u)| \\ &< 2\epsilon. \end{aligned}$$

Hence $F(u,v) \to F(a,b)$ as $u \to a+$ and $v \to b-$. It follows from Corollary 7.10 that f is Henstock integrable to F(a,b) on [a,b].

Theorem 17.8. If a measurable function f has LSRS then f is Henstock integrable on [a,b].

Proof. The proof is similar to that of Theorem 8.12. For completeness, we sketch as follows. If f is Henstock integrable on some subinterval containing x, then we say x is a regular point. It is easy to see that the set of regular points is nonempty. Let Q be the set of all points x which are not regular. Then Q is closed.

Denote by (a_i,b_i), $i = 1,2,\ldots$, the component intervals in $(a,b) - Q$. In view of Lemma 17.7, f is Henstock integrable on $[a_i,b_i]$ for each i. Define U_δ and V_δ as in Definition 17.5. Since they are

VBG^* on [a,b], by the category theorem (Theorem 8.11) there is a portion Q_o of Q such that U_δ and V_δ are $VB^*(Q_o)$. Here Q_o is the intersection of Q and an open interval, say I_o. Let J_o be the smallest interval that contains Q_o, and (c_k,d_k), k = 1,2,..., the component intervals in $J_o - Q_o$. Since U_σ and V_δ are respectively major and minor functions of f, we have

$$V_\delta(v) - V_\delta(u) \le F(v) - F(u) \le U_\delta(v) - U_\delta(u)$$

whenever f is Henstock integrable on [u,v] and F is the primitive of f on [u,v]. Therefore

$$\sum_{k=1}^{\infty} \omega(F;[c_k,d_k]) \quad \text{converges.}$$

Finally, it follows from Theorem 5.8 that f is Henstock integrable on Q_o. Now we apply the Harnack extension (Corollary 7.11) and show that f is Henstock integrable on J_o which is a contradiction. Hence Q is empty and f is Henstock integrable on [a,b].

Theorem 17.9. A measurable function f has LSRS if and only if f is Henstock integrable on [a,b]. Furthermore,

$$\int_a^b f(x)dx = \overline{A} = \underline{A}.$$

Proof. We shall prove only the second part. We recall that

$$\overline{A} = \inf_\delta U_\delta(b) \qquad \text{and} \qquad \underline{A} = \sup_\delta V_\delta(b).$$

Since f has LSRS and is Henstock integrable on [a,b], given $\epsilon > 0$ there is a $\delta(\xi) > 0$ such that for any δ-fine division $D = \{[u,v];\xi\}$ of [a,b] we have

$$\begin{aligned} F(a,b) &\le \sum f(\xi)(v-u) + \epsilon \\ &\le U_\delta(b) + \epsilon \end{aligned}$$

where F(a,b) denotes the integral of f. Therefore $F(a,b) \le \overline{A}$. Similarly, we can prove $\overline{A} \le F(a,b)$. Hence $F(a,b) = \underline{A}$. Following the same argument, we can prove $F(a,b) = \overline{A}$.

In conclusion, the condition of locally small Riemann sums characterizes completely the Henstock integral. We may define the Henstock integral by means of measurable functions satisiying LSRS. Furthermore, the value of the integral is given by $\overline{A}$ or $\underline{A}$.

We remark that if we define |LSRS| as LSRS with δ-fine divisions in Definition 17.1 replaced by δ-fine McShane divisions, then we can show that a measurable function f has |LSRS| if and only if f is McShane integrable on [a,b]. The sufficiency follows as in Theorem 17.2. The necessity is a consequence of the following lemma and Theorem 16.10.

Lemma 17.10. If a measurable function f has |LSRS| then f^+ and f^- have LSRS where $f^+ = \max(f,0)$ and $f^- = \max(-f,0)$.

Proof. We prove for f^+ only. The case for f^- is similar. Given $\epsilon > 0$, there is a $\delta(\xi) > 0$ such that for every $t \in [a,b]$ and any δ-fine McShane division $D = \{[u,v];\xi\}$ of $[r,s] \subset (t - \delta(t), t + \delta(t))$, we have

$$|\sum f(\xi)(v-u)| < \epsilon/2 \quad \text{and} \quad |f(t)(s-r)| < \epsilon/2.$$

Now take any δ-fine division $D_1 = \{[u,v];\xi\}$ of $[r,s] \subset (t-\delta(t), t+\delta(t))$ with $\sum_1$ over D_1. We want to show that

$$|\sum_1 f^+(\xi)(v-u)| < \epsilon.$$

Let D_2 be a division of [r,s] obtained from D_1 in the following way. Let [u,v] be a typical interval in D_1 with the associated point ξ. If $f(\xi) < 0$, replace the associated point ξ by t. Then D_2 thus formed is a δ-fine McShane division. Then with $\sum_2$ over D_2 we obtain

$$\begin{aligned} |\sum_1 f^+(\xi)(v-u)| &\le |\sum_1 f^+(\xi)(v-u) - \sum_2 f(\xi)(v-u)| \\ &\quad + |\sum_2 f(\xi)(v-u)| \\ &\le |f(t)(s-r)| + \epsilon/2 \\ &< \epsilon. \end{aligned}$$

Hence the proof is complete.

18. GLOBALLY SMALL RIEMANN SUMS

Whether a measurable function f is integrable or not depends in some sense on how it behaves on the set of x in which $|f(x)|$ is large, i.e. $|f(x)| > N$ for some N. Therefore we are interested in the following definition.

Definition 18.1. A measurable function f defined on [a,b] is said to have globally small Riemann sums or GSRS if for every $\epsilon > 0$

there exists a positive integer N such that for every $n \geq N$ there is a $\delta_n(\xi) > 0$ and for every δ_n-fine division $D = \{[u,v];\xi\}$ of $[a,b]$ we have

$$\Big|\sum_{|f(\xi)|>n} f(\xi)(v-u)\Big| < \epsilon$$

where the sum is taken over D for which $|f(\xi)| > n$.

By definition, all bounded measurable functions have GSRS since the condition is trivially satisfied.

Theorem 18.2. Let f be Henstock integrable to $F(a,b)$ on $[a,b]$ and $F_n(a,b)$ the integral of f_n on $[a,b]$ where $f_n(x) = f(x)$ when $|f(x)| \leq n$ and 0 otherwise. If $F_n(a,b) \to F(a,b)$ as $n \to \infty$ then f has GSRS.

Proof. Given $\epsilon > 0$ there is a $\delta_n(\xi) > 0$ such that for every δ_n-fine division $D = \{[u,v];\xi\}$ of $[a,b]$ we have

$$|\sum f_n(\xi)(v-u) - F_n(a,b)| < \epsilon,$$

$$|\sum f(\xi)(v-u) - F(a,b)| < \epsilon.$$

Choose N so that whenever $n \geq N$

$$|F_n(a,b) - F(a,b)| < \epsilon.$$

Therefore for $n \geq N$ and δ_n-fine division $D = \{[u,v];\xi\}$ of $[a,b]$ we have

$$\Big|\sum_{|f(\xi)|>n} f(\xi)(v-u)\Big| = |\sum f_n(\xi)(v-u) - \sum f(\xi)(v-u)|$$

$$\leq |\sum f_n(\xi)(v-u) - F_n(a,b)| + |F_n(a,b) - F(a,b)|$$

$$+ |F(a,b) - \sum f(\xi)(v-u)|$$

$$< 3\epsilon$$

Hence f has GSRS.

We remark that the Henstock integrability of f does not imply the above condition $F_n(a,b) \to F(a,b)$ as $n \to \infty$. This can be seen from the following example.

Example 18.3. Let $n = 1,2,\ldots,$ choose x_n such that

$$(x_n - \frac{1}{n+1})/(\frac{1}{n} - x_n) = 1/n.$$

Define $f(x) = n^2$ when $x \in (1/(n+1), x_n)$ and $-n$ when $x \in (x_n, 1/n)$ for $n = 1,2,\ldots,$ otherwise zero. The idea is dividing $(1/(n+1), 1/n)$ into n+1 equal parts and putting f(x) to be n^2 on the first part and $-n$ on the remaining n parts. Then f is improper Riemann integrable on (0,1] and therefore Henstock integrable to 0 on [0,1]. But $F_n(0,1)$, as defined in Theorem 18.2, does not converge to 0 as $n \to \infty$ in view of the nonsymmetric situation. More precisely, writing $m = n^2$ we have

$$|F_m(0,1)| = \sum_{i=n+1}^{m} \frac{i}{(i+1)^2} \geq \frac{1}{2} \sum_{i=n+1}^{m} \frac{1}{i+1}$$

which is divergent as $m \to \infty$.

Theorem 18.4. A measurable function f has GSRS if and only if f is Henstock integrable on [a,b] and $F_n(a,b) \to F(a,b)$ as $n \to \infty$ where $F_n(a,b)$ and F(a,b) are defined as in Theorem 18.2.

Proof. Theorem 18.2 proves the sufficiency. We shall prove only the necessity. Suppose f has GSRS. Note that f_n, as defined in Theorem 18.2, is Henstock integrable on [a,b] for all n. Then for n, m ≥ N and a suitably chosen δ-fine division D = {[u,v];ξ} we have

$$\begin{aligned}|F_n(a,b) - F_m(a,b)| \leq{} & |F_n(a,b) - \sum_{|f(\xi)|\leq n} f(\xi)(v-u)| \\ & + |\sum_{|f(\xi)|\leq m} f(\xi)(v-u) - F_m(a,b)| + |\sum_{|f(\xi)|>n} f(\xi)(v-u)| \\ & + |\sum_{|f(\xi)|>m} f(\xi)(v-u)| \\ <{} & 4\epsilon .\end{aligned}$$

That is, $F_n(a,b)$ converge to a number, say F(a,b), as $n \to \infty$.

Again, for suitably chosen N and δ(ξ) and for every δ-fine division D = {[u,v];ξ} we have

$$\begin{aligned}|\sum f(\xi)(v-u) - F(a,b)| \leq{} & |F(a,b) - F_N(a,b)| \\ & + |F_N(a,b) - \sum_{|f(\xi)|\leq N} f(\xi)(v-u)| + |\sum_{|f(\xi)|>N} f(\xi)(v-u)| \\ <{} & 3\epsilon .\end{aligned}$$

That is, f is Henstock integrable on [a.,b].

In view of Theorem 18.4, we may now define the following.

Definition 18.5. A measurable function f is said to be HL integrable on [a,b] if it has GSRS.

Note that the measurability is assumed in the above definition since it does not follow from GSRS. Note further that the HL integral is a nonabsolute integral intermediate between Lebesgue's and Henstock's. Elementary properties of the HL integral can easily be proved. To show further the technique of proof involving the HL integral, we give a convergence theorem.

Theorem 18.6. If the following conditions are satisfied:

(i) $f_n(x) \to f(x)$ almost everywhere in $[a,b]$ as $n \to \infty$ where each f_n is HL integrable on $[a,b]$,

(ii) f_n has GSRS uniformly in n,

then f is also HL integrable on $[a,b]$ and

$$\int_a^b f_n(x)dx \to \int_a^b f(x)dx \qquad \text{as } n \to \infty.$$

Proof. We may assume that $f_n(x) \to f(x)$ everywhere as $n \to \infty$. Then it follows from (ii) that f has GSRS and is therefore HL integrable on $[a,b]$. It remains to show that the integrals of f_n converge to that of f on $[a,b]$.

Given $\epsilon > 0$, let N and $\delta(\xi) > 0$ be such that for every δ-fine division $D = \{[u,v];\ \xi\}$ of $[a,b]$ we have

$$\Big|\sum_{|f_n(\xi)|>N} f_n(\xi)(v-u)\Big| < \epsilon \qquad \text{for all n.}$$

Here N and $\delta(\xi)$ are independent of n in view of (ii). Then it follows from Egoroff's theorem (Lemma 7.2) that there are an integer n_o and an open set G such that $N|G| < \epsilon$ and whenever $n,\ m \ge n_o$ and $\xi \notin G$

$$|f_{n,N}(\xi) - f_{m,N}(\xi)| < \epsilon$$

where $f_{n,N}(x) = f_n(x)$ when $|f_n(x)| \le N$ and 0 otherwise.

Let $F_n(a,b)$ denote the HL integral of f_n on $[a,b]$. For any fixed $n,\ m \ge n_o$ take $\delta_1(\xi) \le \delta(\xi)$ such that for every δ_1-fine division $D = \{[u,v];\xi\}$ of $[a,b]$ we have

$$|\sum f_i(\xi)(v-u) - F_i(a,b)| < \epsilon \qquad \text{for } i = n,m.$$

Combining all the inequalities above, we have

$$|F_n(a,b) - F_m(a,b)| \le 4\epsilon + |\sum f_{n,N}(\xi)(v-u) - \sum f_{m,N}(\xi)(v-u)|$$

The last term above is less than $\epsilon(b-a) + 2N|G|$ and therefore

$\epsilon(b-a) + 2\epsilon$. Hence $F_n(a,b)$ converges to, say, A on [a,b] as $n \to \infty$.

Next, choose $n \geq n_o$ so that

$$|A - F_n(a,b)| < \epsilon.$$

Write $f_{o,N}(x) = f(x)$ when $|f(x)| \leq N$ and 0 elsewhere. Using the same $\delta_1(\xi)$ above, for every δ_1-fine division $D = \{[u,v];\xi\}$ of [a,b] we have

$$\begin{aligned} |\sum f(\xi)(v-u) - A| &\leq |A - F_n(a,b)| \\ &+ |F_n(a,b) - \sum f_n(\xi)(v-u)| + |\sum_{|f_n(\xi)|>N} f_n(\xi)(v-u)| \\ &+ |\sum_{|f(\xi)|>N} f(\xi)(v-u)| + |\sum f_{o,N}(\xi)(v-u) - \sum f_{n,N}(\xi)(v-u)| \\ &< \epsilon(b-a+6). \end{aligned}$$

That is, f is Henstock integrable to A on [a,b]. Since A is unique, $F_n(a,b)$ converges indeed to the HL integral of f. The proof is complete.

We remark that in Theorem 18.4 the condition $F_n(a,b) \to F(a,b)$ is redundant if f is non-negative. This leads to the following results. Let |GSRS| denote GSRS as in Definition 18.1 with δ_n-fine divisions replaced by δ_n-fine McShane divisions. We recall that McShane divisions are defined in Definition 16.7. It follows from the proof of Theorem 18.4 that a measurable function has |GSRS| if and only if it is absolutely Henstock integrable on [a,b]. Also, if $|f_n(x)| \leq g(x)$ for all x and all n, and if g has |GSRS|, then f_n has |GSRS| uniformly in n. Therefore the dominated convergence theorem for the Lebesgue integral (Theorem 4.3) follows at once from Theorem 18.6. So |GSRS| provides another characterization of the Lebesgue integral.

19. THE DL INTEGRAL

As we mentioned earlier, the Denjoy integral is a countable extension of Lebesgue. Also, the Henstock integral is a countable extension of Riemann. That is, instead of a single δ as in Riemann we take a decreasing null sequence $\{\delta_n\}$ resulting in Henstock. We shall now consider a countable extenion of the RL integral of Section 16.

Definition 19.1. A function f is said to be DL integrable on [a,b] if there exists a number A such that for every $\epsilon > 0$ and $\eta > 0$

and for every infinite subset S of positive integers there exist a positive integer $N \in S$, an open set G and a constant $\delta > 0$ such that $G \supset \{x; |f(x)| > N\}$ and $N|G| < \eta$ and that for every division $D = \{[u,v];\xi\}$ with $0 < v - u < \delta$ and $\xi \in [u,v] - G$ we have

$$\left| \sum_{\xi \notin G} f(\xi)(v-u) - A \right| < \epsilon.$$

Again, it is understood that when $[u,v] \subset G$ the term $f(\xi)(v-u)$ is omitted from the Riemann sum. Otherwise we always choose $\xi \in [u,v] - G$. Here DL stands for Denjoy and Lebesgue.

Theorem 19.2. The DL integral is uniquely determined.

Proof. Suppose there exist two numbers A_1 and A_2 satisfying the defintion. For every $\epsilon > 0$ and $\eta > 0$, taking S to be the set of all positive integers, by definition there exist N_1, G_1 and δ_1 such that $G_1 \supset \{x; |f(x)| > N_1\}$, $N_1|G_1| < \eta$ and that for every division $D = \{[u,v];\xi\}$ with $0 < v - u < \delta_1$ and $\xi \in [u,v] - G_1$ we have

$$\left| \sum_{\xi \notin G_1} f(\xi)(v - u) - A_1 \right| < \epsilon.$$

Since we may vary η, indeed there exist N_i, G_i and δ_i such that the above holds true with N_1, G_1 and δ_1 replaced respectively by N_i, G_i, δ_i for $i = 1,2,\ldots$. Here we may assume that $N_i < N_{i+1}$ and $N_i|G_i| \to 0$ as $i \to \infty$.

Now suppose f is also DL integrable to A_2 on $[a,b]$. Taking $S = \{N_i\}$, then there exist $N = N_i$ for some i, G_o and δ_o such that the above holds true again with N_1, G_1, δ_1 and A_1 replaced respectively by N, G_o, δ_o and A_2. Take N as above, $G = G_i \cup G_o$ and $\delta = \min(\delta_i,\delta_o)$. Then $G \supset \{x; |f(x)| > N\}$ and $N|G| < 2\eta$. For any division $D = \{[u,v];\xi\}$ with $0 < v - u < \delta$ and $\xi \in [u,v] - G$ we have

$$\left| \sum_{\xi \notin G} f(\xi)(v-u) - A_1 \right| \le \left| \sum_{\xi \notin G_i} f(\xi)(v-u) - A_1 \right| + \left| \sum_{\xi \in G-G_i} f(\xi)(v-u) \right|,$$
$$< \epsilon + 2\eta.$$

The above is so by the following argument. If $[u,v] - G$ is non-empty then $f(\xi)(v-u)$ is one of the terms in the sum over $\xi \in [u,v] - G_i$, and if $[u,v] - G$ is empty, then $[u,v] - G_i$ may or may not be empty, the collection of all terms $f(\xi)(v-u)$, when $[u,v] - G_i$ is non-empty, is less than $N|G|$ and therefore less than 2η. Similarly, we can prove that

$$|\sum_{\xi \notin G} f(\xi)(v-u) - A_2| < \epsilon + 2\eta.$$

Since ϵ and η are arbitrary, hence $A_1 = A_2$ and the integral is uniquely defined.

The above technique of proof is standard for the RL and DL integrals. The idea is to change G a litte bit so that the same inequality still holds. For non-negative functions, the DL integral reduces to that of RL and therefore coincides with the Henstock integral. However, in general, this is not true.

Example 19.3. There is a function which is Henstock integrable but not DL integrable. Let

$$f(x) = n \qquad \text{when} \quad x \in (\frac{1}{2}(\frac{1}{n+1} + \frac{1}{n}), \frac{1}{n})$$

$$= -n \quad \text{when } x \in (\frac{1}{n+1}, \frac{1}{2}(\frac{1}{n+1} + \frac{1}{n}))$$

for $n = 1,2,\ldots$ and $f(x) = 0$ elsewhere in $[0,1]$. Obviously, f is Henstock integrable to 0 on $[0,1]$. Note that f is DL integrable on $[\delta,1]$ for any $\delta > 0$ but not on $[0,1]$ since $n|E_n|$ tends to 1 as $n\to\infty$ where $E_n = \{x; |f(x)| > n\}$.

The above example also shows that the DL integral is not closed under Cauchy extension as it is for the Henstock integral (Corollary 7.10).

Example 19.4. There is a function which is DL integrable but not Henstock integrable. Let $f(x) = -(x \log|x|)^{-1}$ when $-1/2 \le x < 0$ or $0 < x \le 1/2$ and $f(0) = 0$. Since f is not Henstock integrable on $[0,1/2]$, it cannot be Henstock integrable on $[-1/2,1/2]$. However it is DL integrable on $[-1/2,1/2]$. In fact, the DL integral here is the Cauchy principal value of the improper Riemann integral. So this example shows that if a function is DL integrable on $[a,b]$ then it is not necessarily so on a subinterval of $[a,b]$.

In what follows, we shall give a different formulation for the DL integral.

Theorem 19.5. If f is DL integrable on $[a,b]$ then the following conditions are satisfied:

(i) $n|E_n| \to 0$ as $n \to \infty$ where $E_n = \{x; |f(x)| > n\}$ and $|E_n|$ denotes the measure of E_n;

(ii) $\int_a^b f_n(x)dx$ converges as $n \to \infty$ where $f_n(x) = f(x)$ when $|f(x)| \le n$ and 0 when $|f(x)| > n$.

Proof. Let E_n and f_n be defined as in (i) and (ii). Given $\epsilon > 0$ and $\eta > 0$ there exist an integer N, an open set G_o and a constant δ_o such that $G_o \supset E_N$ and $N|G_o| < \eta$ and that for any division $D = \{[u,v];\xi\}$ of $[a,b]$ with $v - u < \delta_o$ and $\xi \in [u,v] - G_o$ we have

$$\left| \sum_{\xi \notin G_o} f(\xi)(v-u) - A \right| < \epsilon.$$

Following the proof as in Theorem 16.3 we see that f_N is RL integrable and therefore DL integrable on $[a,b]$. That is, there exist an open set G_N and a constant δ_N such that for every division $D = \{[u,v];\xi\}$ with $o < v - u < \delta_N$ and $\xi \in [u,v] - G_N$ we have

$$\left| \sum_{\xi \notin G_N} f_N(\xi)(v-u) - A_N \right| < \epsilon.$$

Now take N as above, $G = G_o \cup G_N$ and $\delta = \min(\delta_o, \delta_N)$. Then $G \supset E_N$ and $N|G| < 2\eta$. For any division $D = \{[u,v];\xi\}$ with $o < v - u < \delta$ and $\xi \in [u,v] - G$ following the standard argument as in the proof of Theorem 19.2 we have

$$|A - A_N| \le \left|A - \sum_{\xi \notin G} f(\xi)(v-u)\right| + \left| \sum_{\xi \notin G} f_N(\xi)(v-u) - A_N\right|$$

$$< \epsilon + 2\eta + \epsilon + 2\eta.$$

Since we may fix ϵ and vary η, so we may find a subsequence of $\{A_n\}$ which converges to A. Here A_n denotes the integral of f_n on $[a,b]$. Since every subsequence of $\{A_n\}$ has a sub-subsequence that converges to A, hence the sequence $\{A_n\}$ itself converges to A and (ii) holds. Here we have used the fact in the definition of the DL integral that given an infinite subset S of positive integers there exists $N \in S$ such that the corresponding conditions hold.

Similarly, there is a subsequence $\{N_i\}$ of the positive integers such that

$$N_i|E_{N_i}| \le N_i|G_i| \to 0 \text{ as } N_i \to \infty.$$

Again, every subsequence of positive integers has a sub-subsequence that satisfies the above property. Hence $n|E_n| \to 0$ as $n \to \infty$ and (i) holds.

Theorem 19.6. If f satisfies the conditions in Theorem 19.5 then f is DL integrable on [a,b].

Proof. Let A_n be the integral of f_n and $A_n \to A$ as $n \to \infty$. Given $\epsilon > 0$ and $\eta > 0$, for every infinite subset S of positive integers there exists a positive integer $N \in S$ such that $N|E_N| < \eta/2$ and

$$|A_N - A| < \epsilon.$$

Next, choose an open set G_o such that $G_o \supset E_N$ and $N|G_o - E_N| < \eta/2$. Therefore $N|G_o| < \eta$.

Since f_N is bounded, it is RL integrable on [a,b] and therefore DL integrable there. By definition, there exist an open set G_N and a constant δ_N such that $N|G_N| < \eta$ and that for every division $D = \{[u,v];\xi\}$ with $v - u < \delta_N$ and $\xi \in [u,v] - G_N$ we have

$$\Big|\sum_{\xi\notin G_N} f_N(\xi)(v-u) - A_N\Big| < \epsilon.$$

Now choose N as above, $G = G_o \cup G_N$ and $\delta = \delta_N$. Then $G \supset E_N$ and $N|G| < 2\eta$. For any division $D = \{[u,v];\ \xi\}$ with $v - u < \delta$ and $\xi \in \{[u,v];\xi\}$ with $v - u < \delta$ and $\xi \in [u,v] - G$ following the standard argument as in the proof of Theorem 19.2 we have

$$\begin{aligned}\Big|\sum_{\xi\notin G} f(\xi)(v-u) - A\Big| &\le \Big|\sum_{\xi\notin G_N} f_N(\xi)(v-u) - A_N\Big| + |A_N - A| \\ &\quad + \Big|\sum_{\xi\in G-G_N} f_N(\xi)(v-u)\Big| \\ &< 2\epsilon + 2\eta.\end{aligned}$$

Hence the proof is complete.

If f satisfies conditions (i) and (ii) in Theorem 19.5 then we may define its integral to be

$$\int_a^b f(x)dx = \lim_{n\to\infty} \int_a^b f_n(x)dx.$$

This is known as the A-integral in the literature. We have proved that the DL integral and A-integral are equivalent. As an easy consequence of the definition of the DL integral, we have

Theorem 19.7. If f is improper Riemann integrable on (0,1] with the singular point at 0 and f is DL integrable on [0,1] then there is a sequence of positive integers m(n), n = 1,2,..., such that

$$\lim_{n\to\infty} \frac{1}{n} \sum_{i=m(n)}^{n} f\left(\frac{i}{n}\right) = \int_0^1 f(x)dx.$$

For example, a special case is Example 16.5 in which m(n) is taken to be 1 for all n.

We remark that condition (ii) in Theorem 19.5, and again in Theorem 19.6, may be replaced by

(iii) $\int_a^b f^N(x)dx$ converges as $N \to \infty$ where $f^N(x) = f(x)$ when $|f(x)| \le N$, $f^N(x) = N$ when $f(x) > N$ and $-N$ when $f(x) < -N$.

Indeed, let f_n be defined as in Theorem 19.5 (ii). Then

$$\int_a^b f^N = \int_a^b f_N + N|E_N|.$$

In view of (i), therefore (ii) and (iii) are equivalent.

20. NUMERICAL INTEGRATION

We give a quadrature formula for numerical integration. As a motivation, consider a continuous function f on [0,1]. Then the Riemann integral of f can be computed as follows:

$$\int_0^1 f = \lim_{n\to\infty} \sum_{k=1}^{n} \frac{1}{n} f(x_{nk})$$

where $(k-1)/n \le x_{nk} \le k/n$ for k = 1,2,...,n. We may rewrite the above as follows:

$$\int_0^1 f = \lim_{n\to\infty} \sum_{k=1}^{n} a_{nk} f(x_{nk})$$

where $a_{nk} = 1/n$ for $1 \le k \le n$ and 0 otherwise. Here the matrix (a_{nk}) is uniformly regular, that is, it satisfies the following conditions:

(R1) $\lim_{n\to\infty} a_{nk} = 0$ uniformly over k;

(R2) $\lim_{n\to\infty} \sum_{k=1}^{\infty} a_{nk} = 1$;

(R3) $\sup \{ \sum_{k=1}^{\infty} |a_{nk}|;\ n \geq 1\} < +\infty$.

Therefore it is natural to ask the following question. Let (a_{nk}) be a uniformly regular matrix and f a function defined on [0,1]. For convenience, we put $f(x) = f(0)$ when $x < 0$ and $f(x) = f(1)$ when $x > 1$. Also, a_{nk} is non-negative with $a_{nk} = 0$ for $k > m(n)$. Such matrix is said to be non-negative and row-finite. Very often, $m(n) = n$ and $\sum_{k=1}^{m(n)} a_{nk} = 1$. Then write $a_{no} = 0$ and put

$$\sum_{i=0}^{k-1} a_{ni} \leq x_{nk} \leq \sum_{i=1}^{k} a_{ni} \quad \text{for } k = 1,2,\ldots,n.$$

What is the condition on f so that we may compute the integral of f as above ?

Theorem 20.1. Let (a_{nk}) be a non-negative and row-finite matrix satisfying (R1) and (R2). If f is Riemann integrable on [0,1] then

$$\int_0^1 f = \lim_{n\to\infty} \sum_{k=1}^{m(n)} a_{nk} f(x_{nk})$$

where x_{nk} lies between $\sum_{i=0}^{k-1} a_{ni}$ and $\sum_{i=1}^{k} a_{ni}$ for $k = 1,2,\ldots,m(n)$.

Proof. We write

$$s_n = \sum_{k=1}^{m(n)} a_{nk} f(x_{nk}).$$

This may not be a Riemann sum. However we can form a Riemann sum t_n by either adding or subtracting the following term

$$(1 - \sum_{k=1}^{m(n)} a_{nk})\, f(1).$$

In view of (R2), we have

$$\begin{aligned} |s_n - s_p| &\leq |t_n - t_p| + |(1 - \sum_{k=1}^{m(n)} a_{nk}) f(1)| + |(1 - \sum_{k=1}^{m(p)} a_{pk}) f(1)| \\ &\leq |t_n - t_p| + 2\epsilon \end{aligned}$$

for sufficiently large n and p. In view of (R1) and the integrability of f, $\{t_n\}$ is a Cauchy sequence and consequently so is $\{s_n\}$. Hence $\{s_n\}$ and $\{t_n\}$ converge to the same limit which is the integral of f.

We remark that (R3) follows from (R2) when (a_{nk}) is non-negative. Since piecewise continuous functions and functions of bounded variation are Riemann integrable, their integrals may be computed using the above quadrature formula. Some unbounded functions may also have their integrals computed in this way, such as Example 16.5 and Theorem 19.7.

Example 20.2. Let $a_{nk} = 2k/[n(n+1)]$ for $1 \le k \le n$ and 0 for $k > n$. Then

$$\int_0^1 \frac{dx}{\sqrt{x}} = \lim_{n\to\infty} \sum_{k=1}^{n} \frac{2k}{n(n+1)} \left[\sum_{i=1}^{k} \frac{2i}{n(n+1)}\right]^{-1/2}$$

$$= \lim_{n\to\infty} \sum_{k=1}^{n} \frac{2k}{\sqrt{k(k+1)n(n+1)}} .$$

Again, this is an easy consequence of the definition of the RL integral. The above quadrature gives a faster rate of convergence than that of Example 16.5.

Theorem 20.3. Let f and $\hat{f}$ be improper Riemann integrable on (0,1] where $\hat{f}(t) = \sup\{|f(x)|;\ t \le x \le 1\}$. Let (a_{nk}) be a non-negative and row-finite matrix satisfying (R1) and the following condition :

$$\sum_{k=1}^{m(n)} a_{nk} = 1.$$

Suppose that there are positive constants δ and η both independent of n and k with $\delta < 1$ such that

$$a_{nk} \le \eta(x_{nk} - x_{n,k-1})$$

whenever $0 < x_{nk} < \delta$ and $\sum_{i=1}^{k-1} a_{ni} \le x_{nk} \le \sum_{i=1}^{k} a_{ni}$ for $k = 2,3,\ldots, m(n)$. Then we have

$$\int_0^1 f = \lim_{n\to\infty} \sum_{k=2}^{m(n)} a_{nk} f(x_{nk}).$$

Proof. Define $f_B(t) = f(t)$ when $B \le t \le 1$ and $f(B)$ otherwise. Then writing

$$Q_n(f) = \sum_{k=2}^{m(n)} a_{nk} f(x_{nk})$$

we have

$$|Q_n(f) - \int_0^1 f| \le |\int_0^1 f - \int_0^1 f_B| + |\int_0^1 f_B - Q_n(f_B)|$$
$$+ |Q_n(f_B) - Q_n(f)|.$$

It is easy to see that each of the first two terms of the right side of the above inequality is small for sufficiently large B and n. Here we have used Theorem 20.1 For the third term, we have

$$|Q_n(f_B) - Q_n(f)| \le \sum_{x_{nk}<B} a_{nk} |f_B(x_{nk}) - f(x_{nk})|$$
$$\le 2\eta \sum_{x_{nk}<B} (x_{nk} - x_{n,k-1}) \hat{f}(x_{nk})$$
$$\le 2\eta \int_0^B \hat{f}(t)dt$$

which is again small for sufficiently large B. Hence the proof is complete.

Note that the inequalities in the proof may also be used to estimate the error of the quadrature. Geometrically, the condition $a_{nk} \le \eta(x_{nk} - x_{n,k-1})$ says that the spacing of x_{nk} should be in line with the length of the interval in which x_{nk} lies, that is, a_{nk}. For example, when $x_{n,k-1}$ and x_{nk} are the endpoints of the associated interval, the condition is satisfied with $\eta = 1$. It is interesting to point out that numerical integration involving uniformly regular matrices has easy two-dimensional extension, whereas some other methods do not have.

CHAPTER 5 SOME GENERALIZATIONS

21. INTEGRATION IN THE EUCLIDEAN SPACE

There are many generalizations of the Henstock integral. We shall present a few in this chapter. We shall give only brief introduction to each of them and refer to the literature for further details.

The first, and also the one most people are interested in, is the n-dimensional case. We shall present a version and illustrate how the controlled convergence theorem may be proved for this case.

Let E be an interval in the n-dimensional euclidean space, that is, it is the set of all points $x = (x_1, \ldots, x_n)$ with $a_j \le x_j \le b_j$ for $j = 1,2,\ldots,n$. We write $E = [a,b]$ where $a = (a_1, \ldots, a_n)$ and $b = (b_1, \ldots, b_n)$. Assume that a norm has been defined on the n-dimensional euclidean space, for example.

$$\|x\| = (\sum_{j=1}^{n} x_j^2)^{1/2}.$$

An open sphere $S(x,r)$ with centre x and radius r is the set of all y such that $\|x-y\| < r$. Let an interval E be given. A division D of E is a finite collection of interval-point pairs (I,x) with the intervals non-overlapping and their union E. Here x is the associated point of I. We write $D = \{(I,x)\}$. It is said to be δ-fine if for each interval-point pair (I,x) we have $I \subset S(x,\delta(x))$ where x is a vertex of I. If $I = [\alpha,\beta]$ with $\alpha = (\alpha_1, \ldots, \alpha_n)$ and $\beta = (\beta_1, \ldots, \beta_n)$, any point $\gamma = (\gamma_1, \ldots, \gamma_n)$ with $\gamma_j = \alpha_j$ or β_j is called a vertex of I. Note that for each I there are 2^n vertices. The volume of $I = [\alpha,\beta]$ is

$$|I| = \prod_{j=1}^{n} (\beta_j - \alpha_j).$$

Two intervals I_1 and I_2 are non-overlapping if $|I_1 \cap I_2| = 0$.

Definition 21.1. A real number H is the value of the generalized Riemann integral of f over E if given $\epsilon > 0$ there is a positive function $\delta(x)$ such that

$$|(D)\sum f(x)|I| - H| < \epsilon$$

for all δ-fine divisions $D = \{(I,x)\}$ of E.

Obviously, δ-fine divisions exist. The generalized Riemann integral is uniquely determined. We write

$$H = \int_E f(x)dx.$$

If F(E) is the integral of f on E and F(I) that of f on $I \subset E$ then F is called the primitive of f. Here F is an interval function. We may define a corresponding point function as follows. Write $E = [a,b]$ and $x \in E$. Define $F(x) = 0$ when $x_j = a_j$ for at least one j, and $F(x) = F([a,x])$ otherwise. Note that we use F to denote both the interval and point functions. Obviously, if f is generalized Riemann integrable on E then its primitive point function F exists and is continuous. Here the primitive F as an interval function is additive in the sense that if $I = \cup_{i=1}^m I_i$ and I_i are non-overlapping then

$$F(I) = \sum_{i=1}^{m} F(I_i).$$

Conversely, given a point function F defined on E we may define a corresponding interval function as follows. Let $I = [\alpha,\beta]$ with $\alpha = (\alpha_1,\ldots,\alpha_n)$ and $\beta = (\beta_1,\ldots,\beta_n)$. Write $\gamma = (\gamma_1,\ldots,\gamma_n)$ where $\gamma_j = \alpha_j$ or β_j and $n(\gamma)$ denotes the number of terms in γ for which $\gamma_j = \alpha_j$. If F is a function of $x = (x_1,\ldots,x_n)$, i.e., $F(x) = F(x_1,..,x_n)$, then F(I) denotes the sum

$$\sum_{\gamma}(-1)^{n(\gamma)}F(x)$$

where the summation is over all vertices γ. For example, when n = 2,

$$F(I) = F(\beta_1,\beta_2) - F(\alpha_1,\beta_2) + F(\alpha_1,\alpha_2) - F(\beta_1,\alpha_2).$$

Also, when n = 3

$$\begin{aligned} F(I) = {} & F(\beta_1,\beta_2,\beta_3) - F(\alpha_1,\beta_2,\beta_3) + F(\alpha_1,\alpha_2,\beta_3) \\ & - F(\beta_1,\alpha_2,\beta_3) + F(\beta_1,\alpha_2,\alpha_3) - F(\beta_1,\beta_2,\alpha_3) \\ & + F(\alpha_1,\beta_2,\alpha_3) - F(\alpha_1,\alpha_2,\alpha_3). \end{aligned}$$

Note that given a primitive point function, we may recover the primitive interval function as described above. Hence they are mutually convertible.

An elementary set I is either an interval or the union of a finite number of non-overlapping intervals. We need the idea of elementary sets because in what follows we have to consider the difference of two intervals, which is an elementary set and, in general, not an interval. If an interval function is defined for all intervals in E then it is also defined for all elementary sets.

Definition 21.2. Let E be an interval in the n-dimensional euclidean space and $X \subset E$. A function F defined on E is said to be $AC^{**}(X)$ if for every $\epsilon > 0$ there are a $\delta(x) > 0$ and a $\eta > 0$ such that for any two δ-fine partial divisions of E with the associated points in X, namely $D_1 = \{(I_1,x_1)\}$ and $D_2 = \{(I_2,x_2)\}$, with $x_1, x_2 \in X$, satisfying

$$(D_1\backslash D_2)\sum|I| < \eta \quad \text{we have} \quad (D_1\backslash D_2)\sum|F(I)| < \epsilon.$$

We remark that $D_1\backslash D_2 = \{I_1-I_2\}$ in which I_1-I_2 is an elementary set. Here the associated vertices x_1 and x_2 are no longer involved. Also, $D_1\backslash D_2 = \{I\}$ and I denotes a typical elementary set in $D_1\backslash D_2$. Note that D_2 may be void, in which case $D_1\backslash D_2 = D_1$.

To understand the definition of $AC^{**}(X)$, let us refer to the case on the real line. The classical definition of $AC^{*}(X)$ requires two endpoints to belong to X (Definition 6.3). For continuous functions, we may relax the condition to one endpoint (Lemma 6.4 (iii)). In fact, we may further relax the condition to include only δ-fine intervals with associated points in X. If we do so, then we end up with not having enough information about the function on intervals not containing points in X. More precisely, we cannot deduce from this one-point definition that the infinite sum of the oscillations of the function over the contiguous intervals of X is finite, assuming that X is closed. That is why we have to introduce $D_1\backslash D_2$ in Definition 21.2. This is to take care of the oscillation of the function outside X. From another point of view, we require the condition of $AC^{**}(X)$ to hold true for δ-fine McShane divisions (Definition 16.7) and not just ordinary δ-fine divisions. On the real line, the two definitions $AC^{*}(X)$ and $AC^{**}(X)$ are equivalent for the purpose of describing primitive functions. When extended to higher dimensions, $AC^{**}(X)$ provides that extra condition to allow a real-line indpendent proof to go through.

Definiton 21.2a. A function F defined on E is said to be ACG^{**} if $E = \cup_{i=1}^{\infty} X_i$ so that each X_i is closed in E and F is $AC^{**}(X_i)$ for each i. A sequence of functions $\{F_n\}$ is $UACG^{**}$ if $E = \cup_{i=1}^{\infty} X_i$ so that each X_i is closed in E and F_n are $AC^{**}(X_i)$ uniformly in n for each i.

Now we can formulate the controlled convergence theorem as follows.

Theorem 21.3. If the following conditions are satisfied:

(i) $f_n(x) \longrightarrow f(x)$ everywhere in E as $n \to \infty$ where each f_n is generalized Riemann integrable on E;

(ii) the primitives F_n of f_n are $UACG^{**}$,

then f is also generalized Riemann integrable on E and we have

$$\int_E f_n \longrightarrow \int_E f \quad \text{as} \quad n \to \infty.$$

We remark that we could have considered convergence almost everywhere in (i) above. For simplicity, we keep to everywhere. We shall now proceed to prove a series of theorems and finally use them to prove the above theorem. However we shall assume without proof those simple properties of the integral given in Section 3.

Theorem 21.4. Let f_n, $n = 1,2,\ldots$, be generalized Riemann integrable on E with the primitives F_n, $n = 1,2,\ldots$. Suppose $f_n(x) \longrightarrow f(x)$ everywhere in E as $n \to \infty$, and $F_n(x)$ converges to a limit function F(x) for every $x \in E$. Then in order that f is generalized Riemann integrable on E with the primitive F, it is necessary and sufficient that for every $\epsilon > 0$ there exists a function M(x) taking integer values such that for infinitely many $m(x) \geq M(x)$ there is a $\delta(x) > 0$ such that for any δ-fine division $D = \{(I,x)\}$ we have

$$|(D)\sum F_{m(x)}(I) - F(E)| < \epsilon.$$

The proof is identical to that of Corollary 9.4.

Theorem 21.5. Let f be generalized Riemann integrable on E with the primitive F. If F is $AC^{**}(X)$ with X closed in E then f_X and its absolute value $|f_X|$ are both generalized Riemann integrable on E where $f_X(x) = f(x)$ when $x \in X$ and 0 otherwise.

Proof. We shall prove that $f^* = \max(f_X, 0)$ is generalized Riemann integrable on E. If so, then both $f_X = \max(f_X, 0) - \max(-f_X, 0)$ and $|f_X| = \max(f_X, 0) + \max(-f_X, 0)$ are generalized Riemann integrable on E.

Let x be a vertex of I and put $F^*(I) = \max\{F(I), 0\}$ when $x \in X$ and 0 otherwise. Since f is generalized Riemann integrable on E, for every $\epsilon > 0$ there is a $\delta(x) > 0$ such that for any δ-fine division $D = \{(I,x)\}$ of E we have

$$(D)\sum|f(x)|I| - F(I)| < \epsilon.$$

Here we have used Henstock's lemma (Theorem 3.7).

Next, let E_1 be a subinterval in E and define

$$\chi(E_1) = \sup(D)\sum|f(x)|I| - F(I)|$$

where the supremum is taken over all δ-fine divisions D of E_1. Note that χ is superadditive in the sense that if $I = \cup_{i=1}^m I_i$ and I_i non-overlapping then

$$\sum_{i=1}^{m} \chi(I_i) \le \chi(I).$$

Then for any δ-fine division D of E we have

$$|(D)\sum(f^*(x)|I| - F^*(I))| \le (D)\sum \chi(I) \le \epsilon.$$

Since F is $AC^{**}(X)$, we can define

$$H_\delta = \sup(D)\sum F^*(I) \text{ and } H = \inf_\delta H_\delta$$

where the supremum is over all δ-fine divisions $D = \{(I,x)\}$ with $x \in X$ and the union of I in D containing X. Choose an open set $G \supset X$ such that the outer measure $|G-X| < \eta$ where η comes from the definition of $AC^{**}(X)$. Here the outer measure is defined in the same way as in Theorem 5.6 with one dimensional intervals replaced by n-dimensional. Now modify the above $\delta(x)$ if necessary so that $S(x,\delta(x)) \subset G$ when $x \in X$. Fix a δ-fine division D_1 such that

$$H_\delta - \epsilon < (D_1)\sum F^*(I) \le H_\delta.$$

Modify $\delta(x)$ again if necessary so that if D is δ-fine then it is finer than D_1, i.e., any interval in D lies in some interval in D_1. We observe that for any other δ-fine division $D = \{(I,x)\}$ with $x \in X$ and the union of I in D containing X we have

$$(D_1 \backslash D)\sum|I| < \eta$$

which implies

$$(D_1 \backslash D)\sum|F(I)| < \epsilon.$$

Then we have

$$\begin{aligned}|(D)\sum f^*(x)|I| - H_\delta| &\le |(D)\sum\{f^*(x)|I| - F^*(I)\}| \\ &\quad + H_\delta - (D_1)\sum F^*(I) + (D_1 \backslash D)\sum|F(I)| \\ &< 3\epsilon.\end{aligned}$$

Further, there is a $\delta_1(x) > 0$ such that

$$H \le H_{\delta_1} < H+\epsilon.$$

We may assume $\delta(x) \le \delta_1(x)$ for all x. Hence for any δ-fine division D of E we have

$$\begin{aligned}|(D)\sum f^*(x)|I|-H| &\le |(D)\sum f^*(x)|I|-H_{\delta_1}| + H_{\delta_1} - H \\ &< 4\epsilon.\end{aligned}$$

Consequently, f^* is generalized Riemann integrable on E and the proof is complete.

Theorem 21.6. Let $f_{n,X}(x) = f_X(x) = 0$ when $x \notin X$ for all n and X closed. If the following conditions are satisfied:

(i) $f_{n,X}(x) \longrightarrow f_X(x)$ everywhere in E as $n \to \infty$ where each $f_{n,X}$ is generalized Riemann integrable on E;

(ii) the primitives $F_{n,X}$ of $f_{n,X}$ are $AC^{**}(X)$ uniformly in n, then f_X is generalized Riemann integrable on E with the primitive F_X and

$$F_{n,X}(E) \longrightarrow F_X(E) \quad \text{as } n \to \infty.$$

Proof. In view of Theorem 21.4, it is sufficient to prove that

$$F_X(x) = \lim_{n\to\infty} F_{n,X}(x) \quad \text{exists for } x \in E$$

and that the condition in Theorem 21.4 holds with $F_{m(x)}$ and F replaced respectively by $F_{m(x),X}$ and F_X. By (ii), for every $\epsilon > 0$ there exist a

$\delta(x) > 0$ and a $\eta > 0$, both independent of n, such that for any δ-fine partial division D of E with the associated points in X satisfying $(D)\sum|I| < \eta$ we have

$$(D)\sum|F_{n,X}(I)| < \epsilon.$$

By Egoroff's theorem, there is an open set G with $|G| < \eta$ such that

$$|f_n(x)-f_m(x)| < \epsilon \quad \text{for } n,m \ge N \text{ and } x \notin G.$$

Consider the following, in which D is a δ-fine division of $E_1 \subset E$ and $D = D_1 \cup D_2$ so that D_1 contains the intervals with the associated vertices $x \notin G$ and D_2 otherwise,

$$\begin{aligned}|F_{n,X}(E_1)-F_{m,X}(E_1)| &= |(D)\sum\{F_{n,X}(I)-F_{m,X}(I)\}| \\ &\le (D_1)\sum|F_{n,X}(I)-f_{n,X}(x)|I|| \\ &+ (D_1)\sum|F_{m,X}(I)-f_{m,X}(x)|I|| \\ &+ (D_1)\sum|f_{n,X}(x)-f_{m,X}(x)||I| \\ &+ (D_2)\sum|F_{n,X}(I)| + (D_2)\sum|F_{m,X}(I)|.\end{aligned}$$

Since each $f_{n,X}$ is generalized Riemann integrable on E, we can choose a suitable δ-fine division D so that

$$|F_{n,X}(E_1)-F_{m,X}(E_1)| < \epsilon\ (4 + |E|) \quad \text{for } n,m \ge N.$$

In other words, $F_X(x)$ exists for $x \in E$. In fact, for any δ-fine partial division D of E repeating the above process we have

$$|(D)\sum\{F_{n,X}(I)-F_{m,X}(I)\}| < \epsilon \quad \text{for } n,m \ge N.$$

Applying the above result, we can find a subsequence $F_{n(j),X}$ of $F_{n,X}$ such that for any δ-fine partial division D of E we have

$$|(D)\sum\{F_{n(j),X}(I)-F_X(I)\}| < \epsilon 2^{-j} \quad \text{for } j = 1,2,\ldots .$$

Then putting $M(x) = n(1)$, for infinitely many $m(x) = n(j) \ge M(x)$ and for suitable δ-fine divisions D we have

$$|(D)\sum F_{m(x),X}(I) - F_X(E)| \le \sum_{j=1}^{\infty} \epsilon 2^{-j} = \epsilon.$$

The proof is complete.

Theorem 21.7. If the conditions in Theorem 21.6 are satisfied, then for every $\epsilon > 0$ there is an integer N such that for every $n \ge N$ there is a $\delta(x) > 0$ such that for any δ-fine division D of E we have

$$(D)\sum|F_{n,X}(I) - F_X(I)| < \epsilon.$$

Proof. It follows from the first part of the proof of Theorem 21.6 by letting $m \to \infty$.

Theorem 21.8. Let f_n be generalized Riemann integrable on E with the primitive F_n for $n = 1,2,\ldots$. If F_n are $AC^{**}(X)$ uniformly in n, then for every $\epsilon > 0$ there is a $\delta(x) > 0$, independent of n, such that for any δ-fine partial division $D = \{(I,x)\}$ of E with $x \in X$ we have

$$|(D)\sum\{F_{n,X}(I)-F_n(I)\}| < \epsilon.$$

Proof. Since F_n are $AC^{**}(X)$ uniformly in n, for every $\epsilon > 0$ there are a $\delta(x) > 0$ and a $\eta > 0$, both independent of n, such that the rest of the condition for $AC^{**}(X)$ holds. For each n, there is a $\delta_n(x) > 0$ with $\delta_n(x) \le \delta(x)$ such that for any δ_n-fine partial division D of E we have

$$|(D)\sum\{F_n(I)-f_n(x)|I|\}| < \epsilon,$$

$$|(D)\sum\{F_{n,X}(I)-f_{n,X}(x)|I|\}| < \epsilon.$$

Here we have used the fact that $f_{n,X}$ is generalized Riemann integrable on E by Theorem 21.5. Now take any δ-fine partial division D of E with the associated points in X, construct δ_n-fine division of each of I in D and denote the total division by D_1. Split D_1 into D_2 and D_3 so that D_2 contains the intervals with $x \in X$ and D_3 otherwise. Then we obtain

$$\begin{aligned}
|(D)\sum\{F_{n,X}(I)-F_n(I)\}| &= |(D_1)\sum\{F_{n,X}(I)-F_n(I)\}| \\
&\le |(D_2)\sum\{\quad\}| + |(D_3)\sum\{\quad\}| \\
&\le |(D_2)\sum\{F_{n,X}(I) - f_{n,X}(X)|I|\}| + |(D_2)\sum\{f_n(x)|I| - F_n(I)\}| \\
&\quad + |(D_3)\sum F_{n,X}(I)| + |(D_3)\sum F_n(I)|.
\end{aligned}$$

Modify $\delta(x)$ if necessary so that $S(x,\delta(x)) \subset E-X$ when $x \notin X$ and that $(D\backslash D_2)\sum|I| < \eta$ as in the proof of Theorem 21.5. Then $F_{n,X}(I) = 0$ when $x \notin X$. Note that $D_3 = D\backslash D_2$. Therefore $(D_3)\sum F_{n,X}(I) = 0$. Consequently,

$$|(D)\sum\{F_{n,X}(I)-F_n(I)\}| < 3\epsilon.$$

That is, the required condition holds.

Proof of Theorem 21.3. In view of $UACG^{**}$ the sequence $\{F_n\}$ is equicontinuous. Then $\{F_n\}$ is uniformly bounded and by Ascoli's theorem [2;p 336] it contains a uniformly convergent subsequence $\{F_{n(i)}\}$. Therefore

$$F(x) = \lim_{i\to\infty} F_{n(i)}(x) \quad \text{exists for } x \in E$$

and it is continuous on E. It remains to show that F is the primitive of f on E.

We shall verify that the condition in Theorem 21.4 is satisfied. Let $E = \cup_{i=1}^{\infty} X_i$ such that F_n are $AC^{**}(X_i)$ uniformly in n for each i. In view of Theorem 21.5, when $X = X_i$ the functions $f_{n,X}$ are generalized Riemann integrable on E, and in view of Theorem 21.8, $F_{n,X}$ are also $AC^{**}(X)$ uniformly in n. Then, by Theorem 21.7, for every $\epsilon > 0$ and X_i, writing $X = X_i$, there exist an integer $n = n(i,j)$ and a $\delta_n(x) > 0$ such that for any δ_n-fine division D of X we have

$$(D)\sum|F_{n,X}(I)-F_X(I)| < \epsilon 2^{-i-j}.$$

In view of Theorem 21.8 again, we may modify $\delta_n(x)$ if necessary so that for the same D above we have

$$(D)\sum|F_{n,X}(I)-F_n(I)| < \epsilon 2^{-i},$$

$$(D)\sum|F_X(I)-F(I)| < \epsilon 2^{-i}.$$

We may assume for each i that $F_{n(i,j)}$ is a subsequence of $F_{n(i-1,j)}$. Now consider $f_{n(j)} = f_{n(j,j)}$ in place of the original sequence f_n.

Let $Y_1 = X_1$ and $Y_i = X_i - (X_1 \cup \ldots \cup X_{i-1})$ for $i = 2,3,\ldots$. Given $\epsilon > 0$, when $x \in Y_i$ take $M(x) = n(i)$ and $m(x)$ to have values in $\{n(j); j \geq i\}$. For each $m(x)$, put $\delta(x) = \delta_{m(x)}(x)$. Then for any δ-fine division D, writing $m = m(x)$ and $X = X_i$ when $x \in y_i$, we have

$$|F_m(I)-F(I)| \leq |F_m(I)-F_{m,X}(I)| + |F_{m,X}(I)-F_X(I)| + |F\ (I)-F(I)|$$

and consequently,

$$(D)\sum|F_{m(x)}(I)-F(I)| < 3\epsilon.$$

Hence the condition in Theorem 21.4 is satisfied, and we have completed the proof of Theorem 21.3.

We remark that we may define ACG^{**} in Definition 21.2a without requiring X_i to be closed but only measurable. Then Theorem 21.3 still holds true using the new definition. Indeed, the proof may proceed as follows. Suppose F_n are $UAC^{**}(X_i)$ for each i and $\cup_{i=1}^{\infty} X_i = E$. Then there is a closed set $Y_i \subset X_i$ such that F_n are $UAC^{**}(Y_i)$ and $S = E - \cup_{i=1}^{\infty} Y_i$ is of measure zero. In view of $UACG^{**}$, we have $|F(S)| = 0$. Hence the proof goes through as before.

If f is generalized Riemann integrable on E with primitive F then it is easy to verify that F is $AC^{**}(X_N)$ with

$$X_N = \{x; |f(x)| \le N\}.$$

Therefore F is ACG^{**} in the new sense. In other words, ACG^{**} is a property of the primitive and not an additional condition composed on it.

We have demonstrated how the controlled convergence theorem may be proved for the n-dimensional euclidean space. For further development, see comments and references in the bibliography.

22. APPROXIMATELY CONTINUOUS INTEGRALS

One of the important generalizations of the concept of an ordinary limit is that of an approximate limit. Using the approximate limit, we may define approximate derivatives and consequently Burkill's approximately continuous integrals.

First, we recall that the metric density or the density of a set E at a point x is defined to be

$$\lim_{|I|\to 0} \frac{|E \cap I|}{|I|},$$

if it exists, where I denotes an interval containing x. Note that when E is an open interval containing x, the density of E at x is 1.

Definition 22.1. A number A is said to be the approximate limit of a function f at x_o if for every $\epsilon > 0$ there exists a set D of density 1 at x_o such that

$$|f(x) - A| < \epsilon \quad \text{for every } x \in D.$$

We write

$$\lim_{x\to x_o} \text{ap } f(x) = A.$$

So we define a function f to be approximately continuous at x_o if

$$\lim_{x \to x_o} \text{ap } f(x) = f(x_o).$$

The approximate derivative of f at x_o, denoted by AD $f(x_o)$, is

$$\text{AD } f(x_o) = \lim_{x \to x_o} \text{ap } \frac{f(x) - f(x_o)}{x - x_o} .$$

Example 22.2. Let $I_n = (1/(n+1), 1/n)$ for $n = 1,2,\ldots$ and J_n a subinterval in I_n of length $|J_n| = |I_n|/n$. Define $f(x) = 1$ when $x \in J_n$ for all n and $f(x) = 0$ elsewhere in $[0,1]$. Define f by symmetry in $[-1,0]$. Then f is approximately continuous and approximately differentiable at $x = 0$ though not so in the ordinary sense.

Definition 22.3. A collection Δ of closed subintervals of $[a,b]$ is called an approximate full cover of $[a,b]$ if for every $x \in [a,b]$ there exists a measurable set D_x with $x \in D_x$ and of density 1 at x such that $[u,v] \in \Delta$ whenever $u, v \in D_x$ and $u \le x \le v$.

We remark that for a given $\delta(x) > 0$ the family of all $[u,v]$ in δ-fine divisions (sometimes called a full cover) is an approximate full cover, though not conversely. A division of $[a,b]$ given by

$$a = x_o < x_1 < \ldots < x_n = b \text{ and } \{\xi_1, \xi_2, \ldots, \xi_n\}$$

is called a Δ-partition if Δ is an approximate full cover with $[x_{i-1}, x_i]$ coming from Δ, or more precisely,

$$x_{i-1} \le \xi_i \le x_i \text{ and } x_{i-1}, x_i \in D_{\xi_i} \text{ for all i.}$$

Again, we denote a Δ-partition by $\{[u,v];\xi\}$ in which $[u,v]$ represents a typical interval $[x_{i-1}, x_i]$ and ξ its associated point ξ_i. Here $u \le \xi \le v$ and $u,v \in D_\xi$ where D_ξ has density 1 at ξ.

Lemma 22.4. If Δ is an approximate full cover of $[a,b]$ then there exists a Δ-partition of $[a,b]$.

Proof. For $x \in [a,b]$, let D_x be a measurable set with $x \in D_x$ and of density 1 at x such that $[u,v] \in \Delta$ whenever $u,v \in D_x$ and $u \le x \le v$. Choose $\delta(x) > 0$ such that for $0 < t < \delta(x)$.

$$|D_x \cap [x,x+t]| > t/2 \text{ and } |D_x \cap [x-t,x]| > t/2.$$

Denote by E_n the set of all x for which $\delta(x) \geq (b-a)/n$ and for $i = 1,2,\ldots,n$

$$E_{ni} = E_n \cap [a + \frac{i-1}{n}(b-a), a + \frac{i}{n}(b-a)].$$

Suppose $x, z \in E_{ni}$ with $x < z$. Then

$$|D_x \cap [x,z]| > (z-x)/2 \text{ and } |D_z \cap [x,z]| > (z-x)/2.$$

It follows that

$$|z-x| + |D_x \cap D_z \cap [x,z]| \geq |D_x \cap [x,z]| + |D_z \cap [x,z]|$$
$$> |z-x|$$

In other words, $|D_x \cap D_z \cap [x,z]| > 0$ and there exists $y \in D_x \cap D_z \cap [x,z]$ so that the pair $[x,y]$ and $[y,z]$ forms a Δ-partition of $[x,z]$.

We shall now apply the familiar category argument (see, for example, the proof of Theorem 8.12). We say that a point x is regular if it lies in an interval which has a Δ-partition. Then the set Q of all points x not regular is closed. Since every interval complementary to Q has a Δ-partition, the set Q is either perfect or empty. We shall show that it is empty.

Since $[a,b] = \cup_{n=1}^{\infty} \cup_{i=1}^{n} E_{ni}$, apply the category theorem (Theorem 8.11) and we can find a set E_{ni} which is dense in a portion of Q. That is, E_{ni} is dense in $(c,d) \cap Q$ for some (c,d). We may assume that (c,d) is the smallest interval that contains $(c,d) \cap Q$. Then there are points of E_{ni} in (c,d) arbitrarily close to c and d. Select

$$x \in (c,c+\delta(c)) \cap E_{ni} \text{ and } z \in (d-\delta(d),d) \cap E_{ni}.$$

Obviously, $[x,z]$ has a Δ-partition as shown above. So have $[c,x]$ and $[z,d]$ by an identical argument. Hence $[c,d]$ has a Δ-partition which is a contradiction. The proof is complete.

We remark that the technique of inducing a partition of $[a,b]$ into E_{ni} as in the proof above has been used earlier in Lemma 6.17 and Theorem 8.10. In view of Lemma 22.4, the following definition is now meaningful.

Definition 22.5. A function f is said to be AP integrable to A on $[a,b]$ if for every $\epsilon > 0$ there is an approximate full cover Δ of [a,b] such that for every Δ-partition $\{[u,v];\xi\}$ of [a,b] we have

$$|\sum f(\xi)(v-u) - A| < \epsilon.$$

where $\sum$ sums over the Δ-partition $\{[u,v];\xi\}$.

Here AP stands for approximate Perron. Following the proofs of Theorems 3.7 and 5.7, we see that the primitive of an AP integrable function is approximately continuous everywhere in [a,b] and approximately differentiable almost everywhere in [a,b].

Definition 22.6. Let X be closed in [a,b]. A function F is said to be $AC^*_{ap}(X)$ if for every $\epsilon > 0$ there exists a $\eta > 0$ such that for all $\alpha_1 < \beta_1 < \ldots < \beta_p$, points of X, if $\sum_{k=1}^p(\beta_k-\alpha_k) < \eta$ then for every $\lambda \in (0,1)$ there exists $E_k^\lambda \subset [\alpha_k,\beta_k]$ with α_k, $\beta_k \in E_k^\lambda$ and $|E_k^\lambda| > (1-\lambda)(\beta_k-\alpha_k)$ for $1 \le k \le p$ and satisfying

$$\sum_{k=1}^{p} \omega(F;E_k^\lambda) < \epsilon$$

where $\omega(F;E_k^\lambda) = \sup\{|F(x) - F(y)|;\ x,y \in E_k^\lambda\}$. A function is F said to be ACG^*_{ap} if $[a,b] = \cup_{i=1}^{\infty}X_i$ where each X_i is closed and F is $AC^*_{ap}(X_i)$ for each i.

We shall show that if f is AP integrable then its primitive is ACG^*_{ap}. The proof follows along the same line as that for the Henstock integral.

Definition 22.7. Let $X \subset [a,b]$. A function F is said to be $VB^*_{ap}(X)$ if there is a constant M such that for every $\lambda \in (0,1)$ and for any sequence of non-overlapping intervals $\{[a_k,b_k]\}$ with a_k, $b_k \in X$ for all k there exists $E_k^\lambda \subset [a_k,b_k]$ with $a_k,b_k \in E_k^\lambda$ and $|E_k^\lambda| > (1-\lambda)(b_k-a_k)$ for all k and satisfying

$$\sum_k \omega(F;E_k^\lambda) < M.$$

A function F is VBG^*_{ap} if $[a,b] = \cup_{i=1}^{\infty}X_i$ for which each X_i is closed and F is $VB^*_{ap}(X_i)$ for each i.

Lemma 22.8. If f is AP integrable on [a,b] with the primitive F, then F is VBG^*_{ap}.

Proof. Since f is AP integrable, for $\epsilon > 0$ there is an approximate full cover Δ of $[a,b]$ such that for any Δ-partition $\{[u,v];\xi\}$ we have

$$\sum|f(\xi)(v-u) - F(u,v)| < \epsilon.$$

Let D_x be given for each x as in the definition of Δ. For $\lambda \in (0,1)$ there is a $\delta(x) > 0$ such that whenever $0 < t < \delta(x)$

$$|D_x \cap [x,x+t]| > (1-\lambda/2)t, \quad |D_x \cap [x-t,x]| > (1-\lambda/2)t.$$

Denote by X_n the set of all x for which $\delta(x) \geq (b-a)/n$ and $|f(x)| \leq n$. Also, for $i = 1,2,\ldots, n$

$$X_{ni} = X_n \cap [a + \frac{i-1}{n}(b-a), \ a + \frac{i}{n}(b-a)].$$

For any sequence of non-overlapping intervals $\{[a_k,b_k]\}$ with a_k, $b_k \in X_n$ for all k, following the same argument as in the proof of Lemma 22.4 we have

$$|(D_{a_k} \cap D_{b_k}) \cap [a_k,b_k]| > (1-\lambda)(b_k-a_k).$$

Denote by E_k^λ the set on the left side of the above inequality. Replace E_k^λ by $E_k^\lambda \cup \{a_k,b_k\}$ if necessary. Obviously, $E_k^\lambda \subset [a_k,b_k]$ with a_k, $b_k \in E_k^\lambda$ and $|E_k^\lambda| > (1-\lambda)(b_k-a_k)$ for all k. Furthermore, writing $a_k \leq u_k < v_k \leq b_k$ and u_k, $v_k \in E_k^\lambda$ we have

$$\sum_k|F(u_k,v_k)| \leq \sum_k |F(a_k,u_k)| + \sum_k|F(a_k,b_k)| + \sum_k |F(v_k,b_k)|$$

$$< 3\epsilon + 3n(b-a).$$

That is, F is $VB^*_{ap}(X_{ni})$. Since F is approximately continuous on $[a,b]$, F is also $VB^*_{ap}(\bar{X}_{ni})$ where $\bar{X}_{ni}$ denotes the closure of X_{ni}. Consequently, F is VBG^*_{ap}.

Lemma 22.9. Let f be AP integrable on $[a,b]$ with the primitive F and $f_X(x) = f(x)$ when $x \in X$ and zero otherwise. If F is $VB^*_{ap}(X)$ then f_X is absolutely Henstock integrable on $[a,b]$.

Proof. Since F is $VB^*_{ap}(X)$, then F is $VB(X)$, that is,

$$\sup \sum_k |F(a_k,b_k)| < + \infty$$

where the supremum is taken over all non-overlapping intervals $\{[a_k,b_k]\}$ with a_k, $b_k \in X$ for all k. Let $G(a) = F(a)$, $G(b) = F(b)$,

$G(x) = F(x)$ for $x \in X$ and linearly otherwise. Then G is of bounded variation and $G'(x) = AD\,F(x) = f(x)$ almost everywhere in X. Since G' is absolutely Henstock integrable on $[a,b]$, so is f on X. That is, f_X is absolutely Henstock integrable on $[a,b]$.

Lemma 22.10. Let X be closed in $[a,b]$ and (a_k,b_k), $k = 1,2,\dots,$ the component intervals of $(a,b) - X$. An approximately continuous function F is $AC^*_{ap}(X)$ if and only if F is $AC(X)$ and for any $\lambda \in (0,1)$ there is a set $E_k^\lambda \subset [a_k,b_k]$ with $a_k,b_k \in E_k^\lambda$ and $|E_k^\lambda| > (1-\lambda)(b_k-a_k)$ for all k and satisfying

$$\sum_{k=1}^{\infty} \omega(F;E_k^\lambda) < +\infty.$$

Proof. The necessity is obvious provided we can prove the fact that for any $\lambda \in (0,1)$ we can find $E_k^\lambda \subset [a_k,b_k]$ with a_k, $b_k \in E_k^\lambda$ such that $|E_k^\lambda| > (1-\lambda)(b_k-a_k)$ and $\omega(F;E_k^\lambda) < +\infty$. For convenience, we consider $[a,b]$ in place of $[a_k,b_k]$.

Since F is approximately continuous, for $x \in [a,b]$ and $\lambda \in (0,1)$ there is a measurable set $D_x \subset (x-\delta(x),x+\delta(x))$ such that

$$|D_x \cap [x,x+h]| > (1-\lambda)h, \quad |D_x \cap [x-h,x]| > (1-\lambda)h$$

whenever $0 < h < \delta(x)$ and

$$|F(u)-F(v)| < \epsilon \quad \text{for } u,\ v \in D_x \quad \text{and} \quad u \le x \le v.$$

The family $\{D_x;\ x \in [a,b]\}$ defines an approximate full cover Δ of $[a,b]$. Then there is a Δ-partition of $[a,b]$ given by $a = x_0 < x_1 < \dots < x_n = b$ and $\{\xi_1,\dots,\xi_n\}$ with $\xi_1 = a$ and $\xi_n = b$. Let

$$E^\lambda = \bigcup_{k=1}^{n} D_{\xi_k}.$$

It follows that a, $b \in E^\lambda$ and $|E^\lambda| > (1-\lambda)(b-a)$. Furthermore, if u, $v \in E^\lambda$, $u \in [x_{p-1},x_p]$ and $v \in [x_{q-1},x_q]$, then

$$\begin{aligned} |F(v)-F(u)| &\le \sum_{k=p+1}^{q-1} |F(x_k)-F(x_{k-1})| + |F(x_p) - F(u)| + |F(v) - F(x_{q-1})| \\ &\le n\epsilon. \end{aligned}$$

That is, $\omega(F;E^\lambda)$ is finite.

Conversely, suppose the conditions are satisfied. Since F is

AC(X), for every $\epsilon > 0$ there is a $\eta > 0$ such that for all $\alpha_1 < \beta_1 < \ldots < \beta_p$, points of X, if $\sum_{k=1}^{p}(\beta_k - \alpha_k) < \eta$ then

$$\sum_{k=1}^{p} |F(\beta_k) - F(\alpha_k)| < \epsilon.$$

Choose a sufficiently large N such that

$$\sum_{k=N+1}^{\infty} \omega(F;E_k^\lambda) < \epsilon.$$

We may assume the above $\eta < (b_k - a_k)$ for $1 \le k \le N$. Let $\bar{E}_k^\lambda$ denote the union of $X \cap [\alpha_k, \beta_k]$ and $E_n^\lambda \cap [\alpha_k, \beta_k]$ for all $n \ge N+1$. Then it follows that α_k, $\beta_k \in \bar{E}_k^\lambda$ and $|\bar{E}_k^\lambda| > (1-\lambda)(\beta_k - \alpha_k)$. Furthermore, when u, $v \in \bar{E}_k^\lambda$ we can write

$$|F(v)-F(u)| \le |F(\beta_k)-F(\alpha_k)| + |F(u)-F(\alpha_k)| + |F(\beta_k) - F(v)|.$$

Observing that either $u \in X$ or we may decompose $[\alpha_k, u]$ into $[\alpha_k, \alpha_k'] \cup [\alpha_k', u]$ where $\alpha_k' \in X$ and (α_k', u) does not intersect X with $u \in E_n^\lambda$ for some $n \ge N+1$. Consequently,

$$\sum_{k=1}^{p} \omega(F;\bar{E}_k^\lambda) \le 5\epsilon.$$

Hence F is $AC_{ap}^*(X)$.

Theorem 22.11. If f is AP integrable on [a,b] then its primitive F is ACG_{ap}^*.

Proof. Since F is VBG_{ap}^* by Lemma 22.8, F is $VB_{ap}^*(X_i)$ for each closed X_i with $\cup_{i=1}^{\infty} X_i = [a,b]$. In view of Lemma 22.10, it remains to show that F is $AC(X_i)$ for each i.

Write $X = X_i$ and we have

$$F(x) = F_X(x) + \sum_{k=1}^{\infty} F([a,x] \cap [c_k,d_k])$$

where F_X is the primitive of f_X as defined in Lemma 22.9 and (c_k,d_k), $k = 1,2,\ldots$, are the component intervals of $(a,b)-X$. Obviously, F_X is absolutely continuous on [a,b] by Lemma 22.9. It is easy to verify that the second term in the above decomposition of F is AC(X). Hence so is F and the proof is complete.

In fact, ACG^*_{ap} characterizes the primitive of an AP integrable function, as we shall show in the following.

Theorem 22.12. A function f is AP integrable on [a,b] if and only if there is an approximately continuous function F which is ACG^*_{ap} such that AD $F(x) = f(x)$ almost everywhere in [a,b]. Furthermore, the AP integral of f on [a,b] is F(b)−F(a).

Proof. The necessity follows from Theorem 22.11 and the remark after Definition 22.5. To prove the sufficiency, we proceed as in the proof of Theorem 6.12.

First, AD $F(x) = f(x)$ for $x \in [a,b]-S$ where S has measure zero. For $x \in [a,b]-S$ and $\epsilon > 0$, there exists a set D_x with $x \in D_x$ and of density 1 at x such that whenever $u \le x \le v$ and $u, v \in D_x$

$$|F(u,v) - f(x)(v-u)| \le \epsilon|v-u|$$

where $F(u,v) = F(v) - F(u)$.

Next, F is $AC^*_{ap}(X_i)$ for each i where each X_i is closed and $\cup_{i=1}^{\infty} X_i = [a,b]$. Write $Y_1 = X_1$ and $Y_i = X_i - (X_1 \cup \ldots \cup X_{i-1})$ for i = 2,3,... and

$$S_{ij} = \{x \in S \cap Y_i ; j-1 \le |f(x)| < j\}$$

Then there is a $\eta_{ij} < \epsilon 2^{-i-j} j^{-1}$ such that we can choose a sequence of non-overlapping intervals $\{I_k^{ij}\}$ with all endpoints in X_i and satisfying

$$\bigcup_k I_k^{ij} \supset S_{ij} \quad \text{and} \quad \sum_k |I_k^{ij}| < \eta_{ij}$$

so that for every $\lambda \in (0,1)$ there exists $(E_k^{ij})^\lambda \subset I_k^{ij}$ with the endpoints of I_k^{ij} belonging to $(E_k^{ij})^\lambda$ and $|(E_k^{ij})^\lambda| > (1-\lambda)|I_k^{ij}|$ for all k and satisfying

$$\sum_k \omega(F;(E_k^{ij})^\lambda) < \epsilon 2^{-i-j}.$$

Let $E_k^{ij} = \cup_{n=1}^{\infty} (E_k^{ij})^{1/n}$ and when $x \in S_{ij}$, not an endpoint of I_k^{ij}, put $D_x = E_k^{ij} \cup S_{ij}$ where $x \in I_k^{ij}$ for some k. Note that $|E_k^{ij}| = |I_k^{ij}|$ and E_k^{ij} has density 1 at x.

Finally, the set of all endpoints of I_k^{ij} for $i,j,k = 1,2,\ldots$ is countable, say $x_1, x_2, \ldots$. When $x = x_n$, $n = 1,2,\ldots$, choose D_x with $x \in D_x$ and of density 1 at x so that

$$|F(v) - F(u)| < \epsilon 2^{-n} \quad \text{and} \quad |f(x)(v-u)| < \epsilon 2^{-n}$$

whenever $u, v \in D_x$. Hence we have defined D_x for all $x \in [a,b]$.

The family $\{D_x;\ x \in [a,b]\}$ constitutes an approximate full cover Δ. Take any Δ-partition $\{[u,v];\xi\}$ of $[a,b]$. Split the sum $\sum$ over the Δ-partition into three partial sums $\sum_1$, $\sum_2$ and $\sum_3$ in which $\xi \in [a,b] - S$, $\xi = x_n$ for $n = 1,2,\ldots$ and otherwise in S respectively. Then we obtain

$$\begin{aligned}|\textstyle\sum f(\xi)(v-u)-F(a,b)| &\le \textstyle\sum_1 |f(\xi)(v-u)-F(u,v)| + \sum_2 |F(u,v)| + \sum_2 |f(\xi)(v-u)| \\ &\quad + \textstyle\sum_3 |F(u,v)| + \sum_3 |f(\xi)(v-u)| \\ &< \epsilon(b-a) + 2\sum_n \epsilon 2^{-n} + \sum_{i,j} \epsilon 2^{-i-j} + \sum_{i,j} j\eta_{ij} \\ &< \epsilon(b-a) + 4\epsilon.\end{aligned}$$

That is, f is AP integrable on $[a,b]$.

Definition 22.13. A sequence of functions $\{f_n\}$ is said to be approximately control-convergent to f on $[a,b]$ if the following conditions are satisfied:

(i) $f_n(x) \to f(x)$ almost everywhere in $[a,b]$ as $n \to \infty$ where each f_n is AP integrable on $[a,b]$;

(ii) the primitives F_n of f_n are ACG^*_{ap} uniformly in n;

(iii) the primitives F_n of f_n converge to an approximately continuous function F everywhere in $[a,b]$.

Now we state and prove the controlled convergence theorem for the AP integral. The proof follows identically that of Theorem 7.6.

Theorem 22.14. If $\{f_n\}$ is approximately control-convergent to f on $[a,b]$, then f is AP integrable on $[a,b]$ and

$$\lim_{n\to\infty} \int_a^b f_n(x)dx = \int_a^b f(x)dx.$$

Proof. It is easy to see that F being the limit function of F_n is approximately continuous and ACG^*_{ap}. It remains to show that $AD\,F(x) = f(x)$ almost everywhere in $[a,b]$.

Let $X_1, X_2, \ldots$ be the closed sets in the definition of ACG^*_{ap}. Suppose $X = X_i$ and F_n are $AC^*_{ap}(X)$ uniformly in n. For convenience, we may assume $a, b \in X$. If we can show that AD $F(x) = f(x)$ almost everywhere in X, then the proof is complete. As usual, we put $G_n(x) = F_n(x)$ when $x \in X$ and linearly elsewhere in $[a,b]$. Also $G(x) = F(x)$ when $x \in X$ and linearly elsewhere. Since F_n are AC(X) uniformly in n, then G_n are uniformly absolutely continuous on $[a,b]$. Writing $g_n(x) = G'_n(x)$ for almost all x, $g(x) = f(x)$ when $x \in X$ and $G'(x)$ elsewhere, we see that the conditions in Theorem 7.1 are satisfied with f_n and f replaced by g_n and g. Also, the Lebesgue density theorem [3;p 114] states that X being a measurable set has density 1 at almost all x in X. Hence we obtain AD $F(x) = G'(x) = f(x)$ for almost all x in X. The proof is complete.

Note that Theorem 22.12 gives a descriptive definition of the AP integral. To give a descriptive definition of an integral, we determine a class of primitives. For example, for a class of all differentiable functions as primitives, we obtain the Newton integral. To define the (special) Denjoy integral, we use a class of continuous functions which are ACG^* and differentiable almost everywhere. If we replace the ordinary limits by the approximate limits, we define a class of approximately continuous functions which are ACG^*_{ap} and approximately differentiable almost everywhere. The corresponding integral is the AP integral. In other words, given a limit process we have the corresponding continuity and derivatives. If somehow we can define the corresponding generalized absolute continuity then we have an integral. Of course, we assume that we can also prove the uniqueness of the integral. There are many generalizations of this kind. One of them, namely the CP integral, will be presented in the next section. Two more are mentioned in the bibliography.

23. THE CESARO-PERRON INTEGRAL

It is well-known that the Cesaro sum is useful in summing Fourier series. A similar idea in the integral form can also be used to define derivatives. Hence, correspondingly, we may define the Cesaro-Perron or CP integral.

The C-mean of a function f in (x,x+h), denoted by C(f,x,x+h), is

$$\frac{1}{h}\int_x^{x+h} f(t)dt.$$

Here h may take negative values. When $h < 0$, the above defines the C-mean of f in (x+h,x). Note that $C(f,x,y) = C(f,y,x)$. A function f is said to be C-continuous at x if

$$C(f,x,x+h) \to f(x) \text{ as } h \to 0.$$

The C-derivative of f at x denoted by CD f(x) is

$$\lim_{h\to 0} \frac{C(f,x,x+h) - f(x)}{h/2}.$$

It is easy to see that if f has a C-derivative at x then it is C-continuous there. Also, if f is continuous then it is C-continuous.

Definition 23.1. Let X be closed in [a,b]. A function F is said to be $AC_1^*(X)$ if it is Denjoy integrable on an interval containing X and for every $\epsilon > 0$ there is a $\delta > 0$ such that for any finite sequence of non-overlapping intervals $[a_k,b_k]$, $k = 1,2,\ldots,n$, with a_k, $b_k \in X$ and satisfying

$$\sum_{k=1}^{n} |b_k - a_k| < \delta$$

we have

$$\sum_{k=1}^{n} \sup\{|C(F,a_k,x) - F(a_k)|;\ a_k < x < b_k\} < \epsilon,$$

$$\sum_{k=1}^{n} \sup\{|C(F,b_k,x) - F(b_k)|;\ a_k < x < b_k\} < \epsilon.$$

A function F is said to be AC_1G^* on [a,b] if [a,b] is the union of closed sets X_i, $i = 1,2,\ldots$, over each of which F is $AC_1^*(X_i)$.

In the definition of the Henstock integral, we could have considered only those intervals [u,v] for which the associated point ξ is always an endpoint, i.e. either $\xi = u$ or $\xi = v$. In fact, the Henstock integral was first developed using one-sided intervals. However it was found later to be equivalent to the present version (Definition 2.2) which is simpler. As we shall see later, in the case of the CP integral we must consider one-sided intervals. Indeed, we already did so in the definition of $AC_1^*(X)$.

Definition 23.2. A function f is said to be CP integrable on [a,b] if there is a C-continuous function F which is AC_1G^* on [a,b] and such that CD F(x) = f(x) almost everywhere in [a,b].

The CP integral of f on [a,b] is defined to be F(b) − F(a). The uniqueness of the CP integral follows from Theorem 23.6 below.

Now we state and prove the controlled convergence theorem for the CP integral.

Theorem 23.3. If the following three conditions are satisfied:

(i) $f_n(x) \to f(x)$ almost everywhere in [a,b] as $n\to\infty$ where each f_n is CP integrable on [a,b] with primitive F_n;

(ii) F_n are AC_1G^* uniformly in n;

(iii) $F_n(x) \to F(x)$ as $n\to\infty$ pointwise at each $x \in [a,b]$ where F is C-continuous on [a,b],

then f is CP integrable on [a,b] and

$$\int_a^b f_n(x)dx \to \int_a^b f(x)dx \quad \text{as } n \to \infty.$$

Note that if F is $AC_1^*(X)$ then it is AC(X) in view of the following inequality:

$$|F(b_k)-F(a_k)| \le |C(F,a_k,b_k)-F(a_k)| + |C(F,b_k,a_k)-F(b_k)|.$$

Hence the proof of Theorem 23.3 follows identically from that of Theorem 22.14 with AD F(x) replaced by CD F(x), which we shall not elaborate here. However, in place of the Lebesgue density theorem, we apply the following result.

Lemma 23.4. Let X be a closed set in [a,b]. If F is $AC_1^*(X)$ then CD F(x) exists almost everywhere in X.

Proof. Since F is $AC_1^*(X)$, it is also AC(X). As in the proof of Theorem 22.14, AD F(x) exists almost everywhere in X.

We shall define two functions U and V as follows. Let (a_k,b_k), k = 1,2,..., denote the component intervals of (a,b) − X. Define U(x) = V(x) = F(x) when $x \in X$. When $a_k < x < b_k$, define U(x) to be

$$\max\{\sup_{a_k<x<b_k} C(F,a_k,x),\ \sup_{a_k<x<b_k} C(F,b_k,x),\ C(F,a_k,b_k)\}$$

and $V(x)$ to be the minimum of the same expression above. Let c_k denote the midpoint of a_k and b_k. Then U and V are AC(X) and it follows from the definition of $AC_1^*(X)$ that

$$\sum_{k=1}^{\infty} |U(c_k) - U(a_k)| < +\infty, \quad \sum_{k=1}^{\infty} |U(c_k) - U(b_k)| < +\infty,$$

$$\sum_{k=1}^{\infty} |V(c_k) - V(a_k)| < +\infty, \quad \sum_{k=1}^{\infty} |V(c_k) - V(b_k)| < +\infty.$$

Hence U and V are $AC^*(X)$ by Lemma 6.4(ii). In other words, U and V are differentiable almost everywhere in X and we have

$$U'(x) = AD\ F(x) = V'(x)$$

almost everywhere in X.

It is easy to verify that for $x \in X$ and $h > 0$

$$\frac{C(F,x,x+h) - F(x)}{h/2} \le \frac{C(U,x,x+h) - U(x)}{h/2}$$

Taking $h \to 0$, we obtain

$$CD^+F(x) \le CD^+U(x)$$

where CD^+ denotes the right-hand upper C-derivative. Using the standard notations, we can show that

$$CD^-F(x) \le CD^-U(x), \quad CD_+F(x) \ge CD_+V(x),$$

and $$CD_-F(x) \ge CD_-V(x).$$

That is, CD F(x) exists at the point x where $U'(x) = V'(x)$ occurs. Therefore, CD F(x) exists almost everywhere in X and, when it exists, CD F(x) = AD F(x).

Example 23.5. Let $G(0) = 0$ and

$$G(x) = 2x \sin x^{-2} - 2x^{-1}\cos x^{-2} \quad \text{when} \quad 0 < x \le 1.$$

In fact, $G(x)$ is the derivative of $x^2 \sin x^{-2}$. Here G is C-continuous on [0,1] though not continuous at 0. Further, the derivative of G except at 0 is

$$g(x) = 2 \sin x^{-2} - 2x^{-2} \cos x^{-2} - 4x^{-4} \sin x^{-2}.$$

Hence g is CP integrable on [0,1] though not Denjoy integrable there.

Let $f(x) = x^{-4}\sin x^{-2}$ when $0 < x \le 1$ and $f(0) = 0$. Further, let $f_n(x) = f(x)$ when $1/n \le x \le 1$ and $f_n(x) = 0$ when $0 \le x < 1/n$. Then the sequence $\{f_n\}$ satisfies the conditions in Theorem 23.3 and therefore the conclusion of the thereom holds. In fact, f is the last term of g above. Since the first two terms of g are Denjoy integrable on [0,1], so f is CP integrable on [0,1]. Here f is not Denjoy integrable on [0,1].

Consider a δ-fine division $D = \{[u,v];\xi\}$ in which either $\xi = u$ or $\xi = v$. Given a δ-fine division, we can always convert it into the above form since [u,v] can be decomposed into $[u,\xi]$ and $[\xi,v]$ though one of them may be void. We shall define a pair of interval functions on such divisions. Let ξ be the associated point of [u,v] with $\xi = u$ or v. We write

$$\begin{aligned} \phi(F,u,v) &= -2C(F,u,v) + F(u) + F(v) \quad \text{when } \xi = u, \\ &= 2C(F,v,u) - F(v) - F(u) \quad \text{when } \xi = v. \end{aligned}$$

We shall require this function pair in the following theorem. Note again that ϕ is a function of one-sided intervals.

Theorem 23.6. If f is CP integrable on [a,b] with the primitive F, then for every $\epsilon > 0$ there is a $\delta(x) > 0$ such that for any δ-fine division $D = \{[u,v];\xi\}$ of [a,b] with $\xi = u$ or v we have

$$|F(b) - F(a) - \sum\{\phi(F,u,v) + f(\xi)(v-u)\}| < \epsilon.$$

Proof. Suppose f is CP integrable with the primitive F. Then $CD\,F(x) = f(x)$ for all $x \in [a,b] - S$ where S is of measure zero. For each $x \in [a,b] - S$ there exists a $\delta(x) > 0$ such that whenever $0 < |h| < \delta(x)$

$$|C(F,x,x+h) - F(x) - f(x)h/2| < \epsilon|h|/2.$$

Since F is AC_1G^*, then $[a,b] = \cup_i X_i$ such that F is $AC_1^*(X_i)$ for each i. Here each X_i is closed. Let $Y_1 = X_1$, $Y_i = X_i - (X_1 \cup X_2 \cup \ldots \cup X_{i-1})$ for $i = 2,3,\ldots$ and S_{ij} denote the set of all $x \in S \cap Y_i$ such that $j-1 \le |f(x)| < j$. Then there is a $\eta_{ij} < \epsilon 2^{-i-j} j^{-1}$ such that for any sequence of non-overlapping intervals $\{[a_k,b_k]\}$ with $a_k, b_k \in X_i$ for all k and satisfying $\sum_{k=1}^{\infty}(b_k-a_k) < \eta_{ij}$ we have

$$\sum_{k=1}^{\infty} \sup\{|C(F,a_k,x) - F(a_k)|;\ a_k < x < b_k\} < \epsilon 2^{-i-j}$$

$$\sum_{k=1}^{\infty} \sup\{|C(F,b_k,x) - F(b_k)|;\ a_k < x < b_k\} < \epsilon 2^{-i-j}.$$

We choose $\{I_k^{ij}\}$ with all the endpoints in X_i such that

$$\bigcup_{k=1}^{\infty} I_k^{ij} \supset S_{ij} \quad \text{and} \quad \sum_{k=1}^{\infty} |I_k^{ij}| < \eta_{ij}.$$

For $x \in S_{ij}$, not an endpoint of I_k^{ij}, define $\delta(x) > 0$ so that $(x-\delta(x),\ x + \delta(x)) \subset I_k^{ij}$ for some k.

Finally, consider the endpoints of I_k^{ij} for i,j, k = 1,2,... . Label the points as $x_1, x_2, \ldots$. Since F is C-continuous at each x_n there is a $\delta(x_n) > 0$ such that whenever $0 < |h| < \delta(x_n)$

$$|C(F,x_n,x_n+h) - F(x_n)| < \epsilon 2^{-n}, \quad |f(x_n)h| < \epsilon 2^{-n}.$$

Hence we have defined $\delta(x)$ for all $x \in [a,b]$.

Now take any δ-fine division $D = \{[u,v];\xi\}$ with $\xi = u$ or v. Split the sum $\sum$ over D into three partial sums $\sum_1$, $\sum_2$ and $\sum_3$ in which $\xi \in [a,b] - S$, $\xi = x_n$ for n = 1,2,..., and otherwise respectively. Then we obtain

$$\begin{aligned} &|\textstyle\sum f(\xi)(v-u) + \sum \phi(F,u,v) - F(a,b)| \\ &\le \textstyle\sum_1 |f(\xi)(v-u) + \phi(F,u,v) - F(u,v)| \\ &\quad + \textstyle\sum_2 |f(\xi)(v-u)| + \sum_2 |\phi(F,u,v) - F(u,v)| \\ &\quad + \textstyle\sum_3 |f(\xi)(v-u)| + \sum_3 |\phi(F,u,v) - F(u,v)| \end{aligned}$$

Observing that

$$\begin{aligned} \phi(F,u,v) - F(u,v) &= -2C(F,u,v) + 2F(u) \text{ when } \xi = u, \\ &= 2C(F,v,u) - 2F(v) \text{ when } \xi = v, \end{aligned}$$

and making use of all the inequalities above we obtain

$$|\textstyle\sum\{f(\xi)(v-u) + \phi(F,u,v)\} - F(a,b)| < \epsilon\ (b-a) + 5\epsilon.$$

Hence the proof is complete.

Going through the proof above and noting the value 2 in the expression $\phi(F,u,v) - F(u,v)$, we can now understand the reason why we needed h/2 and not h in the definition of CD F(x).

Corollary 23.7. If f is CP integrable on [a,b] and non-negative then f is Henstock integrable on [a,b].

Proof. Let F be the primitive of f and ϕ defined as in Theorem 23.6. We sketch the proof. First, show that F is nondecreasing for non-negative f. Since F is nondecreasing and C-continuous, then it is continuous. Observe that for h > 0.

$$\phi(F,x,x+h) = -\frac{2}{h}\{\int_x^{x+h} F(t)dt - \frac{1}{2}[F(x)+F(x+h)]h\}.$$

and similarly for h < 0. So we may choose sufficiently small h with $v - u = |h|$ such that

$$|\sum\phi(F,u,v)| < \epsilon.$$

In view of Theorem 23.6, f is Henstock integrable on [a,b].

In fact, the necessary conditions in Theorem 23.6 are also sufficient. In other words, the statement there provides a Riemann-type definition for the CP integral. To prove the sufficiency, we proceed as Theorem 8.8 and show that f is CP integrable in the original sense, i.e., the first definition by Burkill using major and minor functions, which is known to be equivalent to Definition 23.2. The proof is straightforward and therefore we shall not elaborate here. For references, see the bibliography.

We may develop a Cesaro-Perron scale of integrals. In place of the C-mean C(f,x,x+h), we consider the C_r-mean

$$C_r(f,x,x+h) = rh^{-r}\int_x^{x+h} (x+h-t)^{r-1} f(t)dt.$$

A function f is C_r-continuous at x if

$$C_r(f,x,x+h) \to f(x) \text{ as } h \to 0.$$

Similarly, we define

$$C_rD\, f(x) = \lim_{h\to 0} \frac{C_r(f,x,x+h) - f(x)}{h/(r+1)}$$

and so on. Consequently, we can define the C_rP integral where C_1P is simply CP. In the same way, we can prove the controlled convergence theorem for the C_rP integral and give a Riemann-type definition.

The above can be further extended to some generalized mean

involving convergence factors, for example,

$$\int_x^{x+h} \nu(x,h,t)f(t)dt, \qquad \int_x^{x+h} f(t)d_tN(x,h;t)$$

where $\nu(x,h,t)$ and $N(x,h;t)$ are suitably defined. The idea is to consider a set of conditions on $\nu(x,h,t)$ or $N(x,h;t)$ so that they resemble the mean $C(f,x,x+h)$ and the usual results on integration remain valid. Then the Cesaro-Perron integral becomes an important and interesting special case. See the bibliography for references.

24. THE KUNUGI INTEGRAL

We shall describe what is known as the ER integral by Kunugi and discuss the space of all ER integrable functions. Here ER stands for espace rangé, i.e., ranked space in French. In what follows, we shall call it the Kunugi integral. In fact, the Kunugi integral as we shall define here is equivalent to the DL or A-integral of Section 19. To motivate, consider the following

$$\lim_{n\to\infty} \int_{X_n} f(t)dt$$

where $X_n \subset X_{n+1}$ and their union is $[a,b]$. When a is the only singular point, we may assume $X_n = [a+1/n, b]$ and take the above limit to be the integral of the function. This is the so-called Cauchy extension. In general, when the singularities are scattered all over $[a,b]$ we can no longer use one and the same sequence $\{X_n\}$ to define the integral. Hence we have to use many different sequences and then show that the integral is independent of the choice of these sequences.

Let E denote the set of all step functions defined on $[a,b]$. We define a neighbourhood in E as follows. Let X be a closed set in $[a,b]$, $\epsilon > 0$ and $f \in E$. A neighbourhood of f, denoted by $V(X,\epsilon;f)$, is the set of all step functions g in E such that $g = f + r$ and r satisfies the following properties:

(α) $|r(x)| < \epsilon$ for all $x \in X$;

(β) $N|E_N(r)| < \epsilon$ for each N where $E_N(r) = \{x;\ |r(x)| > N\}$;

(γ) $\left|\int_a^b r^N(x)dx\right| < \epsilon$ for each N where r^N denotes the truncated function of r, i.e., $r^N(x) = r(x)$ when $|r(x)| \le N$, $r^N(x) = N$ when $r(x) > N$, and $-N$ when $r(x) < -N$. Here N runs over all positive numbers.

A sequence of neighbourhoods $\{V(X_n, \epsilon_n; f_n)\}$ in E is said to be fundamental if $V(X_n, \epsilon_n; f_n) \supset V(X_{n+1}, \epsilon_{n+1}; f_{n+1})$ for each n, $\epsilon_n \to 0$ as $n \to \infty$, and each $G_n = [a,b]-X_n$ satisfies the following condition

(δ) $|G_n| < 2^{-n}$.

For simplicity, we further assume that a fundamental sequence has the properties : $|X_n - X_{n+1}| = 0$ for each n, i.e., X_n is included in X_{n+1} almost everywhere, and ϵ_n converges decreasingly to 0.

A neighbourhood satisfying (δ) is of rank n. We can prove that the space E with the above neighbourhoods is a ranked space. However we shall introduce the Kunugi integral, or what is known as the special ER integral in the literature, without reference to the ranked space.

Lemma 24.1. If $\{V_n\}$ is a fundamental sequence of neighbourhoods in E with $V_n = V(X_n, \epsilon_n; f_n)$, then

(i) $\lim_{n\to\infty} f_n(x)$ exists almost everywhere in $[a,b]$;

(ii) $\lim_{n\to\infty} \int_a^b f_n(x)dx$ exists.

Proof. Since $f_m \in V_n$ for $m > n$, condition (α) gives

$$|f_m(x)-f_n(x)| < \epsilon_n \qquad \text{for } x \in X_n.$$

For every $x \in \cap_{i=n}^{\infty} X_i$ and given $\epsilon > 0$, there is an integer $p > n$ with $2\epsilon_p < \epsilon$ such that whenever $i, k \ge p$ we have

$$|f_i(x)-f_k(x)| < 2\epsilon_p.$$

Hence (i) holds, or more precisely, the limit in (i) exists for all $x \in \cup_{n=1}^{\infty} \cap_{i=n}^{\infty} X_i$. Here we have used condition (δ) so that the convergence in (i) holds true almost everywhere.

Again, since $f_m \in V_n$ for $m > n$, condition (γ) gives

$$\left|\int_a^b \{f_m(x)-f_n(x)\}dx\right| < \epsilon_n.$$

Consequently, (ii) holds.

Definition 24.2. A function f is said to be Kunugi integrable on [a,b] if there exists a fundamental sequence $\{V(X_n, \epsilon_n; f_n)\}$ in E such that $f_n(x)$ converges to f(x) almost everywhere in [a,b] as $n \to \infty$. The integral of f is defined to be

$$\int_a^b f(x)dx = \lim_{n\to\infty} \int_a^b f_n(x)dx.$$

The uniqueness of the Kunugi integral follows from Theorems 24.4 and 24.5. In other words, the integral is defined independent of the choice of fundamental sequences.

The following lemma will be needed later.

Lemma 24.3. Let f and g be measurable functions, f^N the truncated function of f, and $E_N(f) = \{x; |f(x)| > N\}$. Then we have

(i) $|E_N(f+g)| \le |E_{N/2}(f)| + |E_{N/2}(g)|$;

(ii) $\left|\int_a^b [f+g]^{2N} - \int_a^b f^N - \int_a^b g^N\right| \le 2N|E_N(f) \cup E_N(g)|$.

Proof. First, we claim that

$$\{x; |f(x)+g(x)| > N\} \subset \{x; |f(x)| > N/2\} \cup \{x; |g(x)| > N/2\}.$$

Indeed, if $|f(x)+g(x)| > N$ and $|f(x)| \le N/2$ then

$$|g(x)| \ge |f(x)+g(x)| - |f(x)| > N/2.$$

Hence (i) follows.

Considering all cases, namely, $x \in E_N(f) - E_N(g)$, $x \in E_N(g) - E_N(f)$ and $x \in E_N(f) \cap E_N(g)$, we obtain

$$|[f+g]^{2N}(x) - f^N(x) - g^N(x)| \le 2N$$

whenever $x \in E_N(f) \cup E_N(g)$. The above expression is 0 when $x \notin E_N(f) \cup E_N(g)$. Hence (ii) follows by integration.

We recall that the A-integral has been defined in Section 19. A function f is said to be A-integrable on [a,b] if the following conditions are satisfied:

(i) $N|E_N(f)| \to 0$ as $N \to \infty$;

(ii) $\lim_{N\to\infty} \int_b^a f^N(x)dx$ exists

where $E_N(f)$ and f^N are defined as in (β) and (γ) above. Here it is sufficient for N to take only positive integer values.

Theorem 24.4. If f satisfies the conditions in Definition 24.2, then f is A-integrable on [a,b].

Proof. Let $\{V(X_n,\epsilon_n;f_n)\}$ be a fundamental sequence in E, the space of all step functions on [a,b], with $f_n(x)$ converging to f(x) almost everywhere. In view of Lemma 24.3 (i) and using the same notation there, we have

$$N|E_N(f)| \le 2(\frac{N}{2}|E_{N/2}(f-f_n) + \frac{N}{2}|E_{N/2}(f_n)|).$$

For fixed n, since f_n is bounded, the last term on the right side above vanishes for sufficiently large N.

Next, denote by $E_N^*(g)$ the set of all $x \in E_N(g)$ such that

$$\lim_{n\to\infty} f_n(x) = f(x) \text{ exists.}$$

Then we see that

$$E^*_{N/2}(f-f_n) \subset \bigcup_{i=1}^{\infty} \bigcap_{m=i}^{\infty} E^*_{N/2}(f_m-f_n).$$

Since $f_m \in V(X_n,\epsilon_n;f_n)$ for $m > n$, condition (β) gives

$$\begin{aligned}\frac{N}{2}|E_{N/2}(f-f_n)| &= \frac{N}{2}|E^*_{N/2}(f-f_n)| \\ &\le \lim_{i\to\infty} \frac{N}{2}|\bigcap_{m=i}^{\infty} E^*_{N/2}(f_m-f_n)| \\ &\le \limsup_{m\to\infty} \frac{N}{2}|E^*_{N/2}(f_m-f_n)| \\ &< \epsilon_n\end{aligned}$$

Combining the above two inequalities we deduce that $N|E_N(f)| \to 0$ as $N \to \infty$.

To proceed, we obtain from Lemma 24.3 (ii) that for fixed n

$$|\int_a^b [f-f_n]^{2N} - \int_a^b f^N + \int_a^b f_n^N| \le 2N|E_N(f) \cup E_N(f_n)|.$$

For sufficiently large N, $f_n^N = f_n$ and $|E_N(f_n)| = 0$. Re-arranging, we write

$$|\int^b f^N - \int^b f| \le 2N|E(f)| + |\int^b [f-f]^{2N}|.$$

Since $f_m \in V(X_n, \epsilon_n; f_n)$ for $m > n$, condition (γ) gives

$$\left|\int_a^b [f-f_n]^{2N}\right| = \lim_{m\to\infty}\left|\int_a^b [f_m-f_n]^{2N}\right| \le \epsilon_n.$$

We have proved in the first part that $N|E_N(f)| \to 0$ as $N\to\infty$. Therefore

$$\lim_{N\to\infty} \int_a^b f^N = \lim_{n\to\infty} \int_a^b f_n \qquad \text{exists.}$$

That is, f is A-integrable on [a,b].

Theorem 24.5. If f is A-integrable on [a,b] then the conditions of Definition 24.2 are satisfied.

Proof. We shall prove it in four steps. First, we observe in what follows that the index N in the conditions of the A-integral may run through all positive real numbers instead of integers. Define $E_N(f)$ and f^N as usual. Let $0 < r < 1$ and N an integer. Then we have

$$(N+r)|E_{N+r}(f)| \le \left(\frac{N+r}{N}\right)N|E_N(f)|$$

both of which tend to 0 as $N + r \to \infty$. Also, we have

$$\left|\int_a^b f^{N+r} - \int_a^b f\right| \le \left|\int_a^b f^{N+r} - \int_a^b f^N\right| + \left|\int_a^b f^N - \int_a^b f\right|$$

$$\le (N+r)|E_N(f)| + \left|\int_a^b f^N - \int_a^b f\right|.$$

Then each term above tends to 0 as $N + r \to \infty$. Hence the required conditions hold.

Next, we define X_i, ϵ_i, f_i and a neighbourhood $V^*(X_i, \epsilon_i, f_i)$; and further we show that $f_{i+1} \in V^*(X_i, \epsilon_i; f_i)$ for each i. We write for $s > 0$

$$\lambda_s = s|E_s(f)| \quad\text{and}\quad \eta_s = \left|\int_a^b f - \int_a^b f^s\right|.$$

It follows from step one that $\lambda_s \to 0$ and $\eta_s \to 0$ as $s \to \infty$. Put

$$\lambda_N^* = \sup\{\lambda_s;\ s \ge N\},\quad \eta_N^* = \sup\{\eta_s;\ s \ge N\}.$$

Then both sequences above converge decreasingly to 0. Let $\epsilon_i = 2^{-4i}$ and choose an increasing sequence $\{n(i)\}$ of positive integers such that

$$\lambda_{n(i)}^* \le \epsilon_i/8 \quad\text{and}\quad \eta_{n(i)}^* \le \epsilon_i/8.$$

The reason for choosing ϵ_i, $\lambda^*_{n(i)}$ and $\eta^*_{n(i)}$ as such will become clear later. Let

$$X_i = \{x;\ |f(x)| \le n(i)\} \quad \text{and } f_i = f^{n(i)}.$$

Now define $V^*(X_i,\epsilon_i,f_i)$ as a neighbourhood of f_i consisting of all measurable functions g such that $g = f_i + r$ and r satisfies conditions (α), (β) and (γ). We claim that $f_{i+1} \in V^*(X_i,\epsilon_i;f_i)$ for each i. In other words, $f_{i+1}-f_i = r$ and r satisfies conditions (α), (β) and (γ). Condition (α) is trivially satisfied since $r(x) = 0$ when $x \in X_i$. Condition (β) follows from the following

$$N|E_N(f_{i+1}-f_i)| \le N|E_{N+n(i)}(f)| < \epsilon_i$$

for each N. To verify (γ), we see that for every x

$$[f(x)-f^{n(i)}(x)]^{N/2} = f^{n(i)+N/2}(x)-f^{n(i)}(x).$$

It implies that

$$\left|\int_a^b [f-f_i]^{N/2}\right| \le \left|\int_a^b (f^{n(i)+N/2}-f)\right| + \left|\int_a^b (f^{n(i)}-f)\right|$$
$$< 2\eta^*_{n(i)}$$

Also, we have

$$N|E_{N/2}(f-f_i)| \le N|E_{n(i)+N/2}(f)| < 2\lambda^*_{n(i)}.$$

Combining the above inequalities and making use of Lemma 24.3 (ii), we obtain for each N

$$\left|\int_a^b [f_{i+1}-f_i]^N\right| \le \int_a^b [f_{i+1}-f]^{N/2}\Big| + \left|\int_a^b [f_i-f]^{N/2}\right|$$
$$+ N|E_{N/2}(f_{i+1}-f)| + N|E_{N/2}(f_i-f)|$$
$$< 2\eta^*_{n(i+1)}+2\eta^*_{n(i)} + 2\lambda^*_{n(i+1)} + 2\lambda^*_{n(i)}$$
$$\le \epsilon_i.$$

That is, (α), (β) and (γ) are satisfied with $r = f_{i+1}-f_i$ and therefore $f_{i+1} \in V^*(X_i,\epsilon_i;f_i)$. Note that $\{V^*(X_i,\epsilon_i;f_i)\}$ is not the required fundamental sequence since X_i may not be closed and f_i is not a step function. This is the end of step two.

For step three, we shall construct Y_i, δ_i, g_i and a neighbourhood $V(Y_i,\delta_i;g_i)$; and further we establish the connection between

$V(Y_i, \delta_i; g_i)$ and $V^*(X_i, \epsilon_i; f_i)$. Since f is measurable, there is a sequence of step functions $\{\varphi_n\}$ converging to f almost everywhere. By Egoroff's theorem (Lemma 7.2), for each i there is a step function $\varphi_{m(i)}$ such that

$$|\varphi_{m(i)}(x) - f(x)| < \min\{\epsilon_i, \epsilon_i/(b-a)\}$$

for $x \notin G_i$ where G_i is open and $4|G_i| < \epsilon_i/n(i)$.

Since G_i is open, $Y_i^* = [a,b] - \cup_{n=i}^{\infty} G_n$ is closed. There is also a closed set $X_i^* \subset X_i$ such that

$$|X_i - X_i^*| < \epsilon_i/2n(i).$$

Now put $Y_i = Y_i^* \cap X_i^*$. We may assume $Y_i \subset Y_{i+1}$ for each i. Therefore Y_i is closed, $Y_i \subset X_i$, and

$$\begin{aligned} |X_i - Y_i| &\le |X_i - X_i^*| + |X_i - Y_i^*| \\ &< \epsilon_i/n(i). \end{aligned}$$

Here we have used the fact that $X_i - Y_i^* \subset \cup_{n=i}^{\infty} G_n$ and $\sum_{k=i}^{\infty} \epsilon_k/n(k) \le 2\epsilon_i/n(i)$. Finally, put

$$g_i = [\varphi_{m(i)}]^{n(i)},$$

that is, the truncated function of $\varphi_{m(i)}$, and $\delta_i = 2^7 \epsilon_i$. Obviously, $|g_i(x)| \le n(i)$ for all x, and

$$|g_i(x) - f_i(x)| < \min\{\epsilon_i, \epsilon_i/(b-a)\} \qquad \text{for } x \in Y_i.$$

We give the connection between f_i and g_i as follows. Consider

$$N|E_N(f_i - g_i)| \le N|E_N(f_i - g_i) \cap Y_i| + N|E_N(f_i - g_i) \cap Y_i^c|$$

where Y_i^c denotes the complement of Y_i and $Y_i^c = ([a,b] - X_i) \cup (X_i - Y_i)$. If $N > \epsilon_i/(b-a)$ then the set $E_N(f_i - g_i) \cap Y_i$ is empty. If $N \le \epsilon_i/(b-a)$ then

$$N|E_N(f_i - g_i) \cap Y_i| < \epsilon_i.$$

Again, when $N > 2n(i)$, the set $E_N(f_i - g_i)$ is empty since $|f_i(x)| \le n(i)$ and $|g_i(x)| \le n(i)$ for all x. When $N \le 2n(i)$, we have

$$\begin{aligned} N|E_N(f_i - g_i) \cap Y_i^c| &\le 2n(i)|[a,b] - X_i| + 2n(i)|X_i - Y_i| \\ &< 4\epsilon_i. \end{aligned}$$

Here we have used the fact that $n(i)|[a,b] - X_i| = n(i)|E_{n(i)}(f)| < \epsilon_i$.

Combining the three inequalities above we obtain

$$N|E_N(f_i-g_i)| < 5\epsilon_i.$$

Furthermore, using the property that $|g_i(x)-f_i(x)| < \epsilon_i/(b-a)$ for $x \in Y_i$ we obtain

$$\int_a^b |g_i-f_i| < \epsilon_i + 2n(i)|Y_i^c|$$
$$< 5\epsilon_i$$

We have established the connection between f_i and g_i. This is the end of step three.

In the final step, we shall show that $\{V(Y_i,\delta_i;g_i)\}$ is a fundamental sequence such that $g_i(x) \to f(x)$ almost everywhere as $i \to \infty$. By construction, Y_i is closed, $Y_i \subset Y_{i+1}$ for each i, g_i is a step function, and δ_i converges decreasingly to 0. Furthermore,

$$|[a,b]-Y_i| \le |[a,b]-X_i| + |X_i-Y_i|$$
$$< 2^{-i}.$$

That is, condition (δ) is satisfied. It is easy to verify that $g_i(x) \to f(x)$ for all $x \in \cup_{n=1}^{\omega}\cap_{i=n}^{\infty} Y_i$ and the set in which the convergence does not hold is of measure zero.

It remains to prove that if $g \in V_{i+1}$ then $g \in V_i$ where $V_i = V(Y_i,\delta_i;g_i)$. Supppse $g \in V_{i+1}$. That is, conditions (α), (β) and (γ) are satisfied with $r = g - g_{i+1}$. We want to prove that conditions (α), (β) and (γ) are satisfied with $r = g-g_i$. We shall make use of the consequences from steps two and three.

To prove (α), we see that for $x \in Y_i$

$$|g(x)-g_i(x)| \le |g(x)-g_{i+1}(x)| + |g_{i+1}(x)-f_{i+1}(x)|$$
$$+ |f_i(x)-g_i(x)|$$
$$< \delta_{i+1} + \epsilon_{i+1} + \epsilon_i \le \delta_i.$$

To prove (β), we apply Lemma 24.3(i) twice and obtain

$$N|E_N(g-g_i)| \le 4\{\tfrac{N}{4}|E_{N/4}(g-g_{i+1})| + \tfrac{N}{4}|E_{N/4}(g_{i+1}-f_{i+1})|$$
$$+ \tfrac{N}{4}|E_{N/4}(f_{i+1}-f_i)| + \tfrac{N}{4}|E_{N/4}(f_i-g_i)|\}$$
$$< 4\{\delta_{i+1} + 5\epsilon_{i+1} + \epsilon_i + 5\epsilon_i\} \le \delta_i.$$

Again, we apply Lemma 24.3(ii) twice and obtain

$$\begin{aligned}
\left|\int_a^b [g-g_i]^N\right| &\le \left|\int_a^b [g-g_{i+1}]^{N/4}\right| + \left|\int_a^b [g_{i+1}-f_{i+1}]^{N/4}\right| \\
&+ \left|\int_a^b [f_{i+1}-f_i]^{N/4}\right| + \left|\int_a^b [f_i-g_i]^{N/4}\right| \\
&+ 6\{\tfrac{N}{4}|E_{N/4}(g-g_{i+1})| + \tfrac{N}{4}|E_{N/4}(g_{i+1}-f_{i+1})| \\
&+ \tfrac{N}{4}|E_{N/4}(f_{i+1}-f_i)| + \tfrac{N}{4}|E_{N/4}(f_i-g_i)|\} \\
&< \delta_{i+1} + 5\epsilon_{i+1} + \epsilon_i + 5\epsilon_i + 6\{\delta_{i+1} + 5\epsilon_{i+1} + \epsilon_i + 5\epsilon_i\} \\
&\le \delta_i .
\end{aligned}$$

That is, (γ) is satisfied. Note that we have to choose ϵ_i which decreases rapidly, and δ_i which is a large multiple of ϵ_i. The actual sizes of ϵ_i and δ_i are ad hoc. They are chosen so that the proof works, and are not meant to be the best possible.

The final step shows that the conditions of Definition 24.2 are satisfied. Hence the proof is complete.

The above two theorems show that the Kunugi integral is equivalent to the A-integral of Section 19. Since the A-integral is uniquely determined, so is the Kunugi integral and their integral values are the same.

In what follows, we shall discuss some functional analytic properties of the Kunugi integral. Let Q be the space of all Kunugi integrable functions on [a,b]. It is easy to prove that Q is a linear space. We define a norm in Q as follows:

$$\|f\|_Q = \sup_{N>0}\left|\int_a^b f^N(x)\,dx\right| + \sup_{N>0} N|E_N(f)|$$

where f^N and $E_N(f)$ are defined respectively as in conditions (γ) and (β). Note that the supremum above is taken over all positive numbers N. Then it is easy to verify that $\|f\|_Q = 0$ if and only if $f(x) = 0$ almost everywhere, $\|\alpha f\|_Q = |\alpha|\ \|f\|_Q$ and

$$\|f + g\|_Q \le 4\{\|f\|_Q + \|g\|_Q\}.$$

Indeed, we may deduce the last inequality from Lemma 24.3 as follows. Note that

$$\left|\int_a^b (f+g)^{2N} - \int_a^b f^N - \int_a^b g^N\right| \le 2N|E_N(f)| + 2N|E_N(g)|.$$

Re-arranging, we obtain

$$\left|\int_a^b (f+g)^{2N}\right| + 2N|E_{2N}(f+g)|$$

$$\le \left|\int_a^b f^N\right| + \left|\int_a^b g^N\right| + 4N|E_N(f)| + 4N|E_N(g)|.$$

Hence the required inequality follows with a constant 4 on the right side.

Theorem 24.6. The space Q is complete under the norm $\|\cdot\|_Q$.

Proof. Let $\{f_n\}$ be a Cauchy sequence with respect to the given norm. Since

$$|E_N(f_n-f_m)| \le \frac{1}{N}\|f_n-f_m\|_Q$$

then f_n converges in measure to a function say f. More precisely, for every $\delta > 0$ the measure of $\{x; |f_n(x)-f(x)| > \delta\}$ tends to 0 as $n \to \infty$. It is well-known [2; p 163] that there is a subsequence of $\{f_n\}$ which converges to f almost everywhere in [a,b]. Thus it is easy to see that $\|f_n-f\|_Q \to 0$ as $n \to \infty$. It remains to show that $f \in Q$.

In view of Lemma 24.3(i), we write

$$N|E_{2N}(f)| \le N|E_N(f-f_n)| + N|E_N(f_n)|$$

$$\le \|f-f_n\|_Q + N|E_N(f_n)|.$$

Then for $\epsilon > 0$ there is a fixed n such that for sufficiently large N we have $N|E_{2N}(f)| < \epsilon$.

Again, in view of Lemma 24.3(ii) we write

$$\left|\int_a^b f^{2N} - \int_a^b f_n^N\right| \le 2\|f-f_n\|_Q + 2N|E_N(f_n)|.$$

Similarly, the right side of the above inequality is less than ϵ for a fixed n and for sufficiently large N. Since

$$\left|\int_a^b f_m - \int_a^b f_n\right| \le \|f_m-f_n\|_Q \to 0 \text{ as } m,n\to\infty,$$

the second condition in the definition of the A-integral holds. Hence f is A-integrable and $f \in Q$.

Theorem 24.7. Every continuous linear functional T defined on the space Q is of the form

$$T(f) = \int_a^b \alpha f(x)dx$$

for all $f \in Q$ and for some constant α where the integral is in the sense of Kunugi.

Proof. If f is absolutely Henstock integrable on [a,b] then $f \in Q$ and

$$\|f\|_Q \le 2\int_a^b |f(x)|dx.$$

Let L denote the space of all absolutely Henstock integrable functions on [a,b] with norm

$$\|f\|_L = \int_a^b |f(x)|dx.$$

Since T is continuous under $\|\cdot\|_Q$ on Q, it is continuous under $\|\cdot\|_L$ on L. Then it is well-known (Theorem 15.4) that

$$T(f) = \int_a^b f(x)g(x)dx$$

for all $f \in L$ and for some bounded measurable function g. Suppose $f \in Q$. Then $\|f^N - f\|_Q \to 0$ as $N \to \infty$ and

$$\begin{aligned} T(f) &= \lim_{N\to\infty} T(f^N) \\ &= \lim_{N\to\infty} \int_a^b f^N g. \end{aligned}$$

We shall claim that g is a constant function. If so, the fact that $f \in Q$ implies

$$T(f) = \lim_{N\to\infty} \alpha\int_a^b f^N = \alpha\int_a^b f.$$

If not, then there are $\beta < \gamma$ such that

$$\{x; g(x) < \beta\} \quad \text{and} \quad \{x; g(x) > \gamma\}$$

both have positive measures. Without loss of generality we may assume $\beta < 0 < \gamma$. Take a subset X from the first and a subset Y from the second of the above two measurable sets such that $|X| = |Y| \ne 0$. Further, decompose X and Y into pairwise disjoint measurable subsets, namely $X = \cup_{n=1}^{\infty} X_n$ and $Y = \cup_{n=1}^{\infty} Y_n$, such that

$$|X_n| = |Y_n| = 2^{-n}|X| \quad \text{for } n = 1,2,\ldots .$$

Now define

$$f_o(x) = \sum_{n=1}^{\infty} 2^n n^{-1} f_n(x)$$

where $f_n(x) = -1$ when $x \in X_n$, $f_n(x) = 1$ when $x \in Y_n$ and 0 otherwise. We can verify that f_o is A-integrable on [a,b] and therefore $f_o \in Q$.

On the other hand,

$$\lim_{N\to\infty} \int_a^b f_o^N g = \lim_{N\to\infty} \int_a^b |f_o^N g|$$

$$\geq (\gamma+\beta)|X| \sum_{n=1}^{\infty} \frac{1}{n}$$

which leads to a contradiction. Hence g is constant and the proof is complete.

There are many interesting applications of the A-integral to Fourier series by Russian authors. The theory of ranked spaces and the Kunugi integral is an important contribution to real analysis. We shall elaborate further elsewhere when an apportunity arises.

25. A GENERAL THEORY

The Henstock integral begins with δ-fine divisions. Suppose we can isolate the properties of the δ-fine divisions that make the integration work. Then we should be able to define the Henstock integral abstractly. A family of such divisions will be called a division space.

We shall describe briefly the Henstock integral in division spaces. A division space consists of three mathematical objects: a space T, a family $\underline{T}$ of intervals in T, and a collection $\underline{A}$ of families of some interval-point pairs (I,t) satisfying certain conditions. Let T be a given space and $\underline{T}$ a family of intervals in T. For example, T = [a,b] and $\underline{T}$ the family of left-closed intervals [u,v) in [a,b]. A subset E of T is an elementary set if E is an interval or a finite union of mutually disjoint intervals. A division D of E is the family of a finite number of mutually disjoint intervals I with the union E. A subfamily $\underline{T}_1$ of $\underline{T}$ divides E if a division D of E exists with the intervals of D belonging to $\underline{T}_1$, and we say that D comes from $\underline{T}_1$.

Let $\underline{U}$ be a family of some interval-point pairs (I,t). For example, $\underline{U}$ is the family of all ([u,v),t) with $u \le t \le v$. Let $\underline{U}_o$ be a subfamily of $\underline{U}$. Then $\underline{U}_o$ divides E if the corresponding family $\underline{T}_o$ of I divides E, where

$$\underline{T}_o = \{I; (I,t) \in \underline{U}_o\}.$$

For example, $\underline{U}_o$ is the family of all ([u,v),t) such that $t - \delta(t) < u \le t \le v < t + \delta(t)$ for some function $\delta(t) > 0$. Finally, $\underline{A}$ is a collection of families $\underline{U}_o$ in $\underline{U}$. For example, $\underline{U}_o$ is the example given above depending on $\delta(t)$ and $\underline{A}$ is a collection of $\underline{U}_o$ for different $\delta(t)$. In the case of the Henstock integral on an interval [a,b] we begin with divisions having associated points. Then we introduce a $\delta(t) > 0$ so that every δ-fine division divides [a,b]. In order to define the integral, we consider a family of δ-fine divisions for different $\delta(t)$. They are correspondingly $\underline{U}$, $\underline{U}_o$ and $\underline{A}$ here. We consider ([u,v),t) in place of ([u,v],t) so that we may deal with disjoint intervals instead of non-overlapping intervals. The two versions are equivalent.

Definition 25.1. The triple $(T,\underline{T},\underline{A})$ is called a division space if the following properties are satisfied:

(i) For every elementary set E of T, there is $\underline{U} \in \underline{A}$ dividing E.

(ii) If $\underline{U}_1, \underline{U}_2 \in \underline{A}$ both dividing E, then there is $\underline{U}_3 \in \underline{A}$, dividing E, with $\underline{U}_3 \subset \underline{U}_1 \cap \underline{U}_2$.

(iii) If $\underline{U}_o \in A$ divides the union of two disjoint elementary sets E_1 and E_2, then a family of some $(I,t) \in \underline{U}_o$ with $I \subset E_1$ belongs to $\underline{A}$ and divides E_1.

(iv) Given disjoint elementary sets E_1 and E_2, if $\underline{U}_1 \in \underline{A}$ divides E_1 with $I \subset E_1$ for all $(I,t) \in \underline{U}_1$ and $\underline{U}_2 \in \underline{A}$ divides E_2 with $I \subset E_2$ for all $(I,t) \in \underline{U}_2$, then there is $\underline{U}_3 \in A$, dividing $E_1 \cup E_2$, with $\underline{U}_3 \subset \underline{U}_1 \cup \underline{U}_2$.

In the language of δ-fine divisions, (i) says that there is a $\delta(t) > 0$ so that every δ-fine division divides [a,b], which is true by Lemma 2.1. Given $\delta_1(t) > 0$ and $\delta_2(t) > 0$, there is a $\delta_3(t)$, namely, $\min\{\delta_1(t),\delta_2(t)\}$, such that every δ_3-fine division is also δ_1-fine and δ_2-fine. This is (ii). Given a family of δ-fine divisions of [a,b] and a subinterval [c,d] of [a,b], (iii) ensures that there is a subfamily that divides [c,d]. On the other hand, if $\delta_1(t)$ is defined on [a,c] and $\delta_2(t)$ on [c,d], then we can define $\delta_3(t)$ which coincides with $\delta_1(t)$ on [a,c) and with $\delta_2(t)$ on (c,b] and further $\delta_3(c) = \min\{\delta_1(c),\delta_2(c)\}$. That is, given $\delta_1(t)$ and $\delta_2(t)$, there is a $\delta_3(t)$ so that every δ_3-fine division of [a,b] comes from those of [a,c] and [c,b]. This describes (iv). These are the properties required in order to define the integral and to prove some simple properties.

In what follows, h(I,t) denotes a real-valued function of (I,t).

Definition 25.2. A real number H is the value of the generalized Riemann integral of h(I,t) over an elementary set E, relative to $\underline{A}$, if given $\epsilon > 0$ there is $\underline{U}_o \in \underline{A}$ dividing E such that

$$|(D)\sum h(I,t)-H| < \epsilon$$

for all divisions D of E from $\underline{U}_o$ where $(D)\sum$ denotes the sum over D.

Condition (i) shows that the above definition has a meaning, and (ii) shows that the integral, if it exists, is unique and linear in h(I,t). Let E_1 and E_2 be two elementary sets with the union E. Then the other two conditions show that if h(I,t) is generalized Riemann integrable on E then so it is on E_1. Conversely, if h(I,t) is integrable on E_1 and E_2 then it is on E.

To prove convergence theorems, we require the division space to satisfy a further condition, namely, the decomposable property. It is a kind of diagonal process. Note that countable additivity is not required here.

Definition 25.3. A division space $(T,\underline{T},\underline{A})$ is said to be decomposable if the following condition holds:

(v) For all elementary sets E, all sequences $\{\underline{U}_j\} \subset \underline{A}$, each $\underline{U}_j$ dividing E, and all sequences $\{X_j\}$ of mutually disjoint subsets of T, there is $\underline{U}_o \in \underline{A}$ dividing E such that

$$\underline{U}_o[X_j] \subset \underline{U}_j[X_j].$$

for j = 1,2,..., where

$$\underline{U}[X] = \{(I,t);\ (I,t) \in \underline{U},\ t \in X\}.$$

Note that the family of δ-fine divisions of [a,b] forms a decomposable division space with $T = [a,b]$, $\underline{I}$ the family of subintervals in $[a,b]$, and $\underline{A}$ the collection of all δ-fine divisions for different $\delta(t)$.

In what follows, we always assume that $(T,\underline{I},\underline{A})$ is a decomposable division space, on which the generalized Riemann integral is defined. If $h(I,t)$ is integrable over E to $H(E)$, we write

$$H(E) = \int_E h(I,t).$$

The next theorem is Henstock's lemma , and thereafter the monotone convergence theorem.

Theorem 25.4. If $h(I,t)$ is generalized Riemann integrable over an elementary set E, then for every $\epsilon > 0$ there is $\underline{U}_o \in \underline{A}$ such that for all divisions D of E from $\underline{U}_o$ we have

$$(D)\sum|h(I,t)-H(I)| < \epsilon$$

where $H(I)$ is the integral of h over $I \subset E$.

Proof. Since $h(I,t)$ is generalized Riemann integrable over E, given $\epsilon > 0$ there is $\underline{U}_o \in \underline{A}$ such that for all divisions D of E from $\underline{U}_o$ we have

$$|(D)\sum h(I,t)-H(E)| < \epsilon/4.$$

Let P be an elementary subset of E. Definition 25.1 (iii) shows that there is $\underline{U}_1 \subset \underline{U}_0$ such that for all divisions D_1 of E–P from $\underline{U}_1$ we have

$$|(D_1)\sum h(I,t)-H(E-P)| < \epsilon/4.$$

Take any division D_2 of P from $\underline{U}_0$. Then $D_1 \cup D_2$ forms a division of E from $\underline{U}_0$ and

$$|(D_2)\sum h(I,t)-H(P)| \le |(D_1 \cup D_2)\sum h(I,t)-H(E)| + |(D_1)\sum h(I,t)-H(E-P)|$$
$$< \epsilon/2.$$

Since P is arbitrary, given a division D of E from $\underline{U}_0$ we may choose $P_1 \subset D$ such that $h(I,t) \ge 0$ for $(I,t) \in P_1$ and $P_2 \subset D$ such that $h(I,t) < 0$ for $(I,t) \in P_2$. Then

$$(D(\sum|h(I,t)-H(I)| \le |(P_1)\sum\{h(I,t)-H(I)\}| + |(P_2)\sum\{h(I,t)-H(I)\}|$$
$$< \epsilon.$$

The proof is complete.

Theorem 25.5. Let $h_n(I,t)$ be generalized Riemann integrable over E to $H_n(E)$ for $n = 1,2,\ldots$ and $h_n(I,t) \longrightarrow h(I,t)$ as $n \to \infty$ for each (I,t). If $h_n(I,t) \le h_{n+1}(I,t)$ for each (I,t) and each n, $H_n(E)$ is bounded above, and given $\epsilon > 0$ there is $\underline{U}_o \in \underline{A}$ and for each $x \in E$ there is n_o such that for $(I,x) \in \underline{U}_o$ and $n \ge n_o$ we have

$$h(I,x) - h_n(I,x) < \epsilon\, g_o(I,x)$$

where g_o is positive and generalized Riemann integrable over E, then $h(I,t)$ is generalized Riemann integrable over E and

$$\int_E h_n(I,t) \to \int_E h(I,t) \quad \text{as} \quad n \to \infty.$$

Proof. Since $H_n(E) \le H_{n+1}(E)$ for each n and $H_n(E)$ is bounded above, then $H_n(E) \to H(E)$ as $n \to \infty$ for some $H(E)$. Given $\epsilon > 0$, there is an integer N such that

$$H(E) - \epsilon < H_N(E) \le H(E).$$

Since $h_n(I,t)$ is generalized Riemann integrable over E, there exists $\underline{U}_n \in \underline{A}$ such that for all divisions D of E from $\underline{U}_n$ we have

$$|(D)\sum h_n(I,t) - H_n(E)| < \epsilon 2^{-n}.$$

For every $x \in E$, there is $n(x) \ge N$ such that for $(I,x) \in \underline{U}_o$

$$h(I,x) - h_{n(x)}(I,x) < \epsilon g_o(I,x).$$

Since g_o is integrable over E, we may assume that the Riemann sum $(D)\sum g_o(I,x)$ is bounded by a number M for all divisions D of E from $\underline{U}_o$. We may further assume that $\underline{U}_n \subset \underline{U}_o$ for all n.

Let $X_j = \{x \in E;\ n(x) = j\}$ for $j = 1,2,\ldots$. The sets X_j are mutually disjoint. Define $\underline{U} \in \underline{A}$ such that

$$\underline{U}[X_j] \subset \underline{U}_j\ [X_j]$$

for $j = 1,2,\ldots$. This is provided by Definition 25.3. For any division D of E from $\underline{U}$, let p be the smallest value of $n(x)$ for $(I,x) \in D$. Then

$$H(E) - \epsilon < H_p(E) \le (D)\sum H_{n(x)}(I) \le H(E).$$

On the other hand,

$$H(E) - 2\epsilon < (D)\sum h_{n(x)}(I,x) < H(E) + \epsilon.$$

Consequently, we obtain

$$\begin{aligned} H(E)-2\epsilon &< (D)\sum h(I,x) \\ &\le (D)\sum \{h_{n(x)}(I,x) + \epsilon g_o(I,x)\} \\ &\le H(E) + \epsilon + \epsilon M \end{aligned}$$

Hence h is generalized Riemann integrable over E to H(E). The proof is complete.

To understand the condition involving g_o in Theorem 25.5, consider the case when $h_n(I,t) = f_n(t)k(I)$ and $h(I,t) = f(t)k(I)$ for all n. If k(I) defines an additive function in I, then $f_n(t) \to f(t)$ as $n \to \infty$ for each t implies the above condition. However, when $h_n(I,t)$ are functions of (I,t), the condition fills an essential gap in the proof. The condition is needed again in Theorem 25.8 below.

Definition 25.6. A finite or infinite sequence $\{h_n(I,t)\}$ is said to have bounded mixed Riemann sums or BMRS if there is $\underline{U}_o \in \underline{A}$ such that for all divisions D of E from $\underline{U}_o$ we have

$$c_1 \le (D)\sum h_{n(x)}(I,x) \le c_2$$

for all choices n(x) and for some constants c_1 and c_2.

Note that if there is $\underline{U}_o \in \underline{A}$ such that for $(I,t) \in \underline{U}_o$ and for all n

$$g_1(I,t) \le h_n(I,t) \le g_2(I,t)$$

where g_1 and g_2 are generalized Riemann integrable over E, then $\{h_n(I,t)\}$ has BMRS. In particular, the sequence $\{h_n\}$ in Theorem 25.5 has BMRS.

Theorem 25.7. Let $h_1(I,t)$ and $h_2(I,t)$ be generalized Riemann integrable over E. If h_1 and h_2 have BMRS then $\min\{h_1,h_2\}$ and $\max\{h_1,h_2\}$ are generalized Riemann integrable over E.

Proof. Let $H_i(I)$ be the integral of h_i over I for i = 1,2. Then for each (I,t) we can show that

$$\begin{aligned} &|\max\{h_1(I,t),h_2(I,t)\} - \max\{H_1(I),H_2(I)\}| \\ &\le |h_1(I,t)-H_1(I)| + |h_2(I,t)-H_2(I)|. \end{aligned}$$

In view of Lemma 25.4, there is $\underline{U}_i \in \underline{A}$ such that for all divisions D of E from $\underline{U}_i$ we have

$$(D)\sum|h_i(I,t)-H_i(I)| < \epsilon \qquad \text{for } i = 1,2.$$

Since h_1 and h_2 have BMRS, there is $\underline{U}_3 \in \underline{A}$ such that the Riemann sum $(D)\sum\max\{h_1(I,t);h_2(I,t)\}$ is also bounded for all D of E from $\underline{U}_3$. By Definition 25.1(ii), there is $\underline{U}_4 \subset \underline{U}_1 \cap \underline{U}_2 \cap \underline{U}_3$ such that

$$H(E) = \sup(D)\sum\max\{H_1(I),H_2(I)\} \qquad \text{exists}$$

where the supremum is over all divisions D of E from $\underline{U}_4$. We shall show that $\max\{h_1,h_2\}$ is generalized Riemann integrable over E to H(E).

Given $\epsilon > 0$, take a division D_1 of E from $\underline{U}_4$ such that

$$H(E) - \epsilon < (D_1)\sum\max\{H_1(I),H_2(I)\} \leq s.$$

For each $(I,t) \in D_1$, let D_2 be a division of I from $\underline{U}_4$. Then we obtain

$$(D_2)\sum\max\{H_1,H_2\} \geq \max\{H_1(I),H_2(I)\}.$$

Next, choose $\underline{U}_5 \subset \underline{U}_4$ so that every division D from $\underline{U}_5$ is finer than D_1. Hence for any division D of E from $\underline{U}_5$ we have

$$H(E) - 3\epsilon < (D)\sum\max\{h_1,h_2\} < H(E) + 2\epsilon.$$

That is, $\max\{h_1,h_2\}$ is generalized Riemann integrable over E. The case for $\min\{h_1,h_2\}$ is similar.

Theorem 25.8. Let $h_n(I,t)$ be generalized Riemann integrable over E for $n = 1,2,\ldots,$ and $h_n(I,t) \to h(I,t)$ as $n \to \infty$ for each (I,t). If $\{h_n\}$ has BMRS, and given $\epsilon > 0$ there is $\underline{U}_o \in \underline{A}$ and for each $t \in E$ there is n_o such that for $(I,t) \in \underline{U}_o$ and $m,n \geq n_o$ we have

$$|h_m(I,t)-h_n(I,t)| < \epsilon g_o(I,t)$$

where g_o is positive and generalized Riemann integrable over E, then h is generalized Riemann integrable over E and

$$\int_E h_n(I,t) \to \int_E h(I,t) \qquad \text{as } n \to \infty.$$

Proof. Let $k_{m,n} = \min\{h_i;\ m \leq i \leq n\}$ and $k_{m,n}(I,t) \to k_m(I,t)$ as $n \to \infty$. Since $\{h_n\}$ has BMRS, the function $k_{m,n}$ is generalized Riemann integrable over E, by Theorem 25.7. Note that $k_{m,n}(I,t) \geq k_{m,n+1}(I,t)$ for each (I,t) and each n. Also, the mixed Riemann sums are bounded below and the corresponding condition involving g_o holds. Then it follows from the monotone convergence theorem (Theorem 25.5) that k_m is generalized Riemann integrable over E and

$$\int_E k_{m,n}(I,t) \to \int_E k_m(I,t) \quad \text{as } n \to \infty.$$

Note again that $k_m(I,t) \le k_{m+1}(I,t)$ for each (I,t) and each m. Also, the mixed Riemann sums are bounded above. Again, h is generalized Riemann integrable over E and $\int_E k_m(I,t) \to \int_E h(I,t)$ as $m \to \infty$.

Combining the above results, we have

$$\int_E h = \lim_{m\to\infty} \int_E k_m \le \liminf_{n\to\infty} \int_E h_n.$$

Similarly, we can obtain

$$\int_E h \ge \limsup_{n\to\infty} \int_E h_n.$$

Hence the required result follows.

In view of the remark after Definition 25.6, Theorem 25.8 corresponds to the dominated convergence theorem. The next step is to give a countable extension of the above theorem. We conjecture that some further condition on the division space is required in order to achieve the extension.

BIBLIOGRAPHY

The references are listed in chronological order in each section below.

A. GENERAL

A short history of Henstock integration appears in [11]. The book by Saks [1] remains a useful reference to nonabsolute integration. Recent books include Kurzweil [7], McLeod [8], McShane [9], Henstock [11], and also Djvarsheishvili [5] on Fourier series, Schwabik [6] on differential equations, and Thomson [10] on differentiation. Among older books, [3] contains a chapter on Denjoy integrals and [2] a chapter on the Perron integral. For a good reference to real analysis, see [4].

[1] S Saks, Theory of the integral, 2nd ed, Warsaw 1937.

[2] E J McShane, Integration, Princeton Univ Press 1944.

[3] R L Jeffery, The theory of functions of a real variable, Univ Toronto Press 1951.

[4] A Zygmund, Trigonometric series I and II, Cambridge Univ Press 1977.

[5] V G Chelidze and A G Djvarsheishvili, Theory of the Denjoy integral and some of its applications (in Russian), Tbilisi 1978; English ed, World Scientific 1989.

[6] S Schwabik, M Turdy, O Vejvoda, Differential and integral equations, D Reidel 1979.

[7] J Kurzweil, Nichtabsolut Konvergente Integrale, Teubner, Leipzig, 1980.

[8] R M McLeod, The generalized Riemann integral, Carus Math Monographs 20(1980), MAA.

[9] E J McShane, Unified integration, Academic Press 1984.

[10] B Thomson, Real functions, Springer-Verlag LN 1170(1985).

[11] R Henstock, Lectures on the theory of integration, World Scientific 1988.

B. THE CONTROLLED CONVERGENCE THEOREM ET AL

A version of the controlled convergence theorem was first proved by Djvarsheishvili [12] and later independently by Lee and Chew [14]. An improvement (Corollary 7.7) was given by Liao [18]. For other related results, see [15,16,17,20]. The idea of using one endpoint to characterize continuous functions which are $AC^*(X)$ as stated in Lemma 6.4 (iii) was suggested by Chew. It simplifies many proofs in chapter 2. The proof that the primitive of a Henstock integrable function is ACG^* is faulty in [13]. It has been corrected in Lemma 6.19. For other proofs, see [19,21].

[12] A G Djvarsheishvili, On a sequence of integrals in the sense of Denjoy, Akad Nauk Gruzin SSR Trudy Mat Inst Rajmadze **18**(1951) 221-236. MR 14-628.

[13] P Y Lee and Wittaya Naak-in, A direct proof that Henstock and Denjoy integrals are equivalent, Bull Malaysian Math Soc (2)**5**(1982) 43-47.

[14] P Y Lee and T S Chew, A better convergence theorem for Henstock integrals, Bull London Math Soc **17**(1985) 557-564.

[15] P Y Lee and T S Chew, A Riesz-type definition of the Denjoy integral, Real Analysis Exchange **11**(1985/86) 221-227.

[16] P Y Lee and T S Chew, On convergence theorems for the nonabsolute integrals, Bull Austral Math Soc **34**(1986) 133-140.

[17] P Y Lee and T S Chew, A short proof of the controlled convergence theorem for Henstock integrals, Bull London Math Soc **19**(1987) 60-62.

[18] K C Liao, A refinement of the controlled convergence theorem for Henstock integrals, SEA Bull Math **11**(1987) 49-51.

[19] R Gordon, Equivalence of the generalized Riemann and restricted Denjoy integrals, Real Analysis Exchange **12**(1986/87) 551-574.

[20] G Q Liu, The measurability of $\delta(\xi)$ in Henstock integration, Real Analysis Exchange **13**(1987/88) 446-450.

[21] D F Xu and S P Lu, Henstock integrals and Lusin's condition (N), Real Analysis Exchange **13**(1987/88) 451-453.

C. FUNCTIONAL ANALYSIS

The β-space (Definition 11.2) was first developed by Sargent [24]. The fact that the Denjoy space is a β-space and of the first category can also be found in [24]. The Banach dual of the Denjoy space (Theorem 12.7) was first proved by Alexiewicz [22], and the Köthe dual (Theorem 12.8) by Sargent [23]. The representation of additive functionals on the Denjoy space is due to Chew [30,31]. Theorem 14.8 is due to Drewnowski and Orlicz [27], though in a different form. The presentation in Sections 14 and 15 proceeding from L_∞ and L to the Denjoy space is that of Chew. Similar results hold for the L_p space [31]. The best paper on the representation of additive functionals in author's view is still [27]. There are further results including those involving abstract spaces. See, for example, [26,28,29]. The scale of spaces given in Example 11.9 is taken from [25].

[22] A Alexiewicz, Linear functionals on Denjoy integrable functions, Coll Math 1(1948) 289-293.

[23] W L C Sargent, On the integrability of a product, J London Math Soc 23(1948) 28-34.

[24] W L C Sargent, On some theorems of Hahn, Banach and Steinhaus, J London Math Soc 28(1953) 438-451.

[25] J C Burkill and F W Gehring, A scale of integrals from Lebesgue's to Denjoy's, Quart J Math Oxford (2)4(1953) 210-220.

[26] N A Friedman and A E Tong, On additive operators, Canad J Math 20(1968) 79-87.

[27] L Drewnowski and W Orlicz, Continuity and representation of orthogonally additive functionals, Bull Acad Polon Sci, Ser sci math, astronom et phys 17(1969) 647-653.

[28] J Batt, Nonlinear integral operators on C(S,E), Studia Math 48(1973) 145-177.

[29] M Marcus and V J Mizel, A Radon-Nikodym theorem for functionals, J Functional Analysis 23(1976) 285-309.

[30] T S Chew, On nonlinear integrals, Proc Analysis Conf Singapore 1986, North-Holland 1988, pp 57-62.

[31] T S Chew, Nonlinear Henstock-Kurzweil integrals and representation theorems, SEA Bull Math 12(1988) 97-108.

D. OTHER RIEMANN-TYPE INTEGRALS

The Riemann-type integrals presented in Chapter 4 are mainly the work of Ding Chuan Song and his school [35,37,39]. For the McShane integral, see [9]. The locally small Riemann sums was first defined by Schurle [33,34]. The presentation in Section 17 is due to Lu. The definition used there is equivalent to the original by Schurle. Theorem 20.3 is due to Lagare [36]. Several papers on Riemann-type integrals written in Chinese, not listed here, were authored by Ding Chuan-Song, Lu Shi-Pan, Ma Zhen-Min, Li Bao-Ling, and Ye Guo-Ju. There are some other interesting papers by C F Osgood, O Shisha and others. See, for example, [32] and several papers that follow. A recent paper is Jarnik and Kurzweil [38].

[32] J T Lewis and O Shisha, The generalized Riemann, simple, dominated and improper integrals, J Approx Theory **38**(1983) 192-199.

[33] A W Schurle, A function is Perron integrable if it has locally small Riemann sums, J Austral Math Soc (Series A) **41**(1986) 224-232.

[34] A W Schurle, A new property equivalent to Lebesgue integrability, Proc Amer Math Soc **96**(1986) 103-106.

[35] C S Ding and P Y Lee, On absolutely Henstock integrable functions, Real Analysis Exchange **12**(1986/87) 524-529.

[36] E M Lagare, Improper Riemann integrals and uniformly regular matrices, SEA Bull Math **11**(1987) 23-26.

[37] C S Ding, Q Zhao, and P Y Lee, A Riemann-type definition of Kunugi's integral, Bull Fac Sci Ibaraki Univ (Series A) **20**(1988) 1-4.

[38] J Jarnik and J Kurzweil, A new and more powerful concept of the PU-integral, Czechoslovak Math J **38**(113)(1988) 8-48.

[39] S P Lu, T S Chew, and P Y Lee, On integrals involving Riemann sums, submitted.

E. GENERALIZATIONS

For Henstock integration in the euclidean space, see [11,42]. A proof of the controlled convergence theorem that is real-line independent can be found in [43]. The idea in Theorem 21.8 is due to Chew, which leads to the definition of $AC^{**}(X)$ and the fact that the primitive of a generalized Riemann integrable function is ACG^{**}. To prove further results involving derivatives and using Vitali's covering theorem, we need regularity of intervals. For example, Chew uses it to show that Henstock and Denjoy integrals are equivalent for the n-dimensional space. One drawback for higher dimensions is that so far we can prove either Fubini's theorem or a version of the divergence theorem but not both within the same system. For Fubini's theorem, see [11]. Two versions of the divergence theorem appeared in [40,41], and in both cases some kind of regularity of intervals is assumed.

[40] J Mawhin, Generalized multiple Perron integrals and the Green-Goursat theorem for differentiable vector fields, Czechoslovak Math J 31(1981) 614-632.

[41] W F Pfeffer, The divergence theorem, Trans Amer Math Soc 295(1986) 665-685.

[42] K M Ostaszewski, Henstock integration in the plane, AMS Memoirs 353(1986).

[43] P Y Lee, Generalized convergence theorems for Denjoy-Perron integrals, Real Analysis Symposium, Coleraine 1988.

The AP integral was first defined by Burkill [44] in terms of major and minor functions, which is equivalent to the Riemann-type definition (Definition 22.5) due to Henstock [60;p 223]. For the equivalent definitions and many related results, see [47] and the references therein. The controlled convergence theorem for the AP integral was proved by Soeparna [48]. For further generalizations, see [45, 46].

[44] J C Burkill, The approximately continuous Perron integral, Math Z 34(1931) 270-278.

[45] Y Kubota, An integral of the Denjoy type I, II and III, Proc Japan Acad 40(1964) 713-713; 42(1966) 737-742; 43(1967) 441-444.

[46] D N Sarkhel and A K De, The proximally continuous integrals, J Austral Math Soc (Series A) 31(1981) 26-45.

[47] P S Bullen, The Burkill approximately continuous integral, J

Austral Math Soc (series A) 35(1983) 236-253.

[48] Soeparna D, The controlled convergence theorem for the approximately continuous integral of Burkill, Proc Analysis Conf Singapore 1986, North-Holland 1988, pp 63-68.

The CP integral was first defined by Burkill [49] using major and minor functions, which is equivalent to the descriptive definition (Definition 23.2) due to Sargent [51]. It was extended to the Cesaro-Perron scale, namely, the C_rP integral by Burkill [50] with its descriptive definition given by Sargent [51]. There is a defect in [51] concerning the proof that the two definitions are equivalent. This was corrected by Verblunsky [53]. A Riemann-type definition of the C_rP integral was given by Cross [54] and the controlled convergence theorem for the CP integral proved by Mugalov [52]. The above can be further extended to some generalized mean involving convergence factors. These have been studied by Jeffery, Miller and Henstock in a series of papers appearing between 1945 and 1960.

[49] J C Burkill, The Cesaro-Perron integral, Proc London Math Soc (2)34(1932) 314-322.

[50] J C Burkill, The Cesaro-Perron scale of integration, Proc London Math Soc (2)39(1935) 541-552.

[51] W L C Sargent, A descriptive definition of Cesaro-Perron integrals, Proc London Math Soc (2)47(1941) 212-247.

[52] A G Mugalov, On the limits under the integral sign in the sense of Cesaro-Perron (in Russian), Izv Akad Nauk Azerbaidzan SSR, Fiz Tehn Mat Nauk (1967) 33-41. MR 35#5557.

[53] S Verblunsky, On a descriptive definition of Cesaro-Perron integrals, J London Math Soc (2)3(1971) 326-333.

[54] G Cross, Generalized integrals as limits of Riemann-like sums, Real Analysis Exchange 13(1987/88) 390-403.

The ER integral was first defined by Kunugi [55]. The version presented in Section 24 is that of Nakanishi [58]. The functional analytic properties (Theorems 24.6 and 24.7) are due to Amemiya and Ando [57]. The equivalence with the A-integral was first proved by Amemiya and Ando [57], and also by Nakanishi [58]. A version on abstract measure space can be found in [56]. A recent article is [59] from which one can trace a long list of articles on the ranked space and the ER integral by the Japanese authors, including T Ikegami, Y Nagakura, and M Washihara. For applications to Fourier series, see references in [57]. The subject of ranked spaces and the Kunugi integral is little known outside Japan.

[55] K Kunugi, Application de la méthode des espaces rangés à la theorie de l'intégration I, Proc Japan Acad 32(1956) 215-220.

[56] H Okano, Sur les Intégrales (E.R.) et ses Applications, Osaka Math J 11(1959) 187-212.

[57] I Amemiya and T Ando, On the class of functions integrable in a certain generalized sense, J Fac Sci Hokkaido Univ 18(1965) 128-140.

[58] S Nakanishi, On generalized integrals I, II and III, Proc Japan Acad 44(1968), 133-138, 225-280, 904-909.

[59] S Nakanishi, Integration of ranked vector space valued functions, Math Japonica 33(1988) 105-128.

A readable account of the general theory of Henstock integration can be found in [61]. A rich source of materials and ideas is provided in [60]. For interested reader, consult the references in [60,61].

[60] R Henstock, Linear analysis, Butterworths 1968.

[61] P Muldowney, A general theory of integration in function spaces, Pitman Research Notes in Math 153, Longmans 1987.

INDEX